AF786332

RESEARCH AND PERSPECTIVES IN NEUROSCIENCES

Fondation Ipsen

Editor

Yves Christen, Fondation Ipsen, Paris (France).

Editorial Board

Albert Aguayo, McGill University, Montreal (Canada).
Philippe Ascher, Ecole Normale Supérieure, Paris (France).
Alain Berthoz, Collège de France, CNRS UPR 2, Paris (France).
Jean-Marie Besson, INSERM U 161, Paris (France).
Emilio Bizzi, Massachusetts Institute of Technology, Boston (USA).
Anders Björklund, University of Lund, Lund (Sweden).
Ira Black, University of Medicine & Dentistry of New Jersey, Piscataway (USA).
Floyd Bloom, Scripps Clinic and Research Foundation, La Jolla (USA).
Joël Bockaert, Centre CNRS-INSERM de Pharmacologie Endocrinologie, Montpellier (France).
Pierre Buser, Institut des Neurosciences, Paris (France).
Jean-Pierre Changeux, Collège de France, Institut Pasteur, Paris (France).
Carl Cotman, University of California, Irvine (USA).
Stephen Dunnett, University of Cambridge, Cambridge (UK).
George Fink, Medical Research Council, Edinburgh (UK).
Fred Gage, Salk Institute, La Jolla (USA).
Jacques Glowinski, Collège de France, Paris (France).
Claude Kordon, INSERM U 159, Paris (France).
Michel Lacour, CNRS URA 372, Marseille (France).
Michel Le Moal, INSERM U 259, Bordeaux (France).
Gary Lynch, University of California, Irvine (USA).
Brenda Milner, McGill University, Montreal (Canada).
John Olney, Washington University Medical School, Saint Louis (USA).
Alain Privat, INSERM U 336, Montpellier (France).
Allen Roses, Duke University Medical Center, Durham (USA).
Constantino Sotelo, INSERM U 106, Paris (France).
Jean-Didier Vincent, Institut Alfred Fessard, CNRS, Gif-sur-Yvette (France).
Bruno Will, Centre de Neurochimie du CNRS/INSERM U 44 Strasbourg (France).

Springer

Berlin
Heidelberg
New York
Barcelona
Budapest
Hong Kong
London
Milan
Paris
Santa Clara
Singapore
Tokyo

A.M. Galaburda Y. Christen (Eds.)

Normal and Abnormal Development of the Cortex

With 47 Figures and 4 Tables

Springer

Galaburda, A.M., M.D.
Beth Israel Deaconess Medical Center
East Campus
330 Brookline Avenue
Boston, MA 02215
USA

Christen, Y., Ph.D.
Fondation IPSEN
24, rue Erlanger
75781 Paris, Cédex 16
France

ISSN 0945-6082
ISBN-13: 978-3-642-64590-7 Springer-Verlag Berlin Heidelberg New York

Library of Congress Cataloging-in-Publication Data.
Normal and abnormal development of the cortex / A.M. Galaburda, Y. Christen (eds.). p. cm. – (Research and perspectives in neurosciences) Includes bibliographical references and index.
ISBN-13: 978-3-642-64590-7 (softcover : alk. paper) Cerebral cortex–Growth. 2. Cerebral cortex–Abnormalities. 3. Developmental disabilities – Physiological aspects. I. Galaburda, Albert M., 1948 –. II. Christen, Yves. III. Series. QP 383.N67 1997 612.8'25–dc21

This work is subject to copyright. All rights are reserved, whether the whole or part of the material is concerned, specifically the rights of translation, reprinting, reuse of illustrations, recitation, broadcasting, reproduction on microfilm or in any other way, and storage in data banks. Duplication of this publication or parts thereof is permitted only under the provisions of the German Copyright Law of September 9, 1965, in its current version, and permission for use must always be obtained from Springer-Verlag. Violations are liable for prosecution under the German Copyright Law.

ISBN-13: 978-3-642-64590-7 e-ISBN-13: 978-3-642-60861-2
DOI: 10.1007/978-3-642-60861-2

© Springer-Verlag Berlin Heidelberg 1997

Softcover reprint of the hardcover 1st edition 1997

The use of general descriptive names, registered names, trademarks, etc., in this publication does not imply, even in the absence of a specific statement, that such names are exempt from the relevant protective laws and regulations and therefore free for general use.

Product Liability: The publishers cannot guarantee the accuracy of any information about dosage and application contained in this book. In every individual case the user must check such information by consulting the relevant literature.

Production: PRO EDIT GmbH, D-69126 Heidelberg

Cover design: Design & Production, D-69121 Heidelberg

Typesetting: Mitterweger Werksatz GmbH, Plankstadt

SPIN: 10576126 27/3136 – 5 4 3 2 1 0 – Printed on acid-free paper

Preface

In keeping with IPSEN's growing interest in the neurosciences, the Fondation organized a meeting dealing with the topic of normal and abnormal cortical development, which took place in Paris, on October 14, 1996. This was the 7th of a series of colloquia devoted to the neurosciences. The topics covered included pre-migratory events controlling cell proliferation, neuronal migration, early cortical plasticity including cell death and development of axonal connections, and genetic and epigenetic effects on cortical teratogenesis.

It is clear that the ultimate structure and functional capacity of the cerebral cortex are determined, at least in part, by developmental events taking place very early, even before neuronal migration. Caviness and colleagues suggest that the regulation of the G_1 portion of the cell cycle may be an important mechanism for determining neuronal class and pre-migratory positional encoding in the pseudostratified periventricular germinal zones. Similarly, Kennedy and Dehay emphasize the role of pre-migratory cell cycle kinetics in the cytoarchitectonic differentiation of the cortex, but also consider some post-migratory events important to cortical differentiation. Among the post-migratory events controlling cortical differentiation are these thalamo-cortical connections, and some of the rules governing this thalamo-cortical interaction was reviewed by Dennis O'Leary at the meeting.

Three papers deal with the topic of neuronal migration. The paper by Hatten and collaborators reviews the role of astrotactin and its cellular and molecular properties. The molecule is linked to chromosome 1 in both mouse and human genomes, and mutations involving the coding region in the human genome lead to abnormalities of brain development. Mutations involving another molecule, the GIRK2 potassium channel, are also associated with defects in migration. As we learn from Sotelo and Soriano, Cajal-Retzius cells appear to be important for the transformation of astrocytes into radial glia cells, a step that is required for cell migration. Pasko Rakic suggests that similar defects of neuronal migration may result from disruption of any of a series of steps involved in the determination of cell position, which include dynamic changes in cell organelles, the action of specific cell recognition receptors, the formation of adhesion molecules, or the activation of specific ion channels. The recognition of such molecular events opens the door to new research on the pathogenesis of cerebral malformation and eventually pharmacologic interventions in appropriate patient populations.

Post-migrational events include the development of neural connections and regulation of cell numbers by cell death. Innocenti tells us that studies of cortical connections must ultimately explain mapping functions, amplification of signals, and transformation along the temporal dimension. The steps used during the development of cortical connections include guidance factors affecting the leading processes of axonal growth, combined with exuberant growth with subsequent selection. As with cell proliferation and neuronal migration, both intrinsic and environmental factors determine the final structure. Finlay and Niederer's work seems to show that, in contrast to the recent emphasis on input for the understanding of cortical differentiation, a given area of cortex will not accept any and all types on inputs, and inputs will not search for alternative targets in an indiscriminate and unlimited fashion. Corticofugal pathways, however, appear to exhibit substantially greater flexibility. Remarkable flexibility is indeed illustrated by the damaged developing cortex. The work by Galaburda and colleagues further indicates that cortical malformations appearing spontaneously in mutant mouse strains, as well as those induced by cortical damage in newborn rats, are associated with changes in neuronal size and connectivity at sites far removed from the area of anomaly. Although the spread of changes is widespread, it appears to follow anatomical connections present during the time of injury. Such systematic changes may help explain the circumscribed cognitive anomalies associated with the malformations.

Christopher Walsh's work with retroviral labeling of proliferating neurons indicates that later in corticogenesis there is broad dispersion of clonally related neurons. Can this phenomenon explain widerspread, yet still specific effects of mutations on cerebral cortical development? Disorders of neuronal migration are giving in to genetic analysis. We learn about work regarding two malformations: periventricular heterotopia and double cortex syndrome. The paper by Evrard and Gressens reviews modern tools for the analysis of genetic and environmental determinants of cortical malformations in humans and the multiple factors that result in malformations, including drugs like alcohol and cocaine. Excitatory agents and trophic factors are involved in normal development but may lead to destructive effects in the presence of ischemia/anoxia or intrauterine infections. Some of these toxic effects may be prevented by molecules that interfere with the actions of the excitotoxins. Finally, Kristina Wisniewski and colleagues report on a study comparing the distribution of three molecules, Bcl-2, Bax, and c-Jun – molecules thought to regulate neuronal survival – in Down Syndrome and in normal brains. An altered distribution in Down syndrome is insinuated as a factor in abnormal brain development in this genetic disorder.

The papers selected for this volume represent only a small fraction of the work being carried out presently on brain development in general, cortical development in particular, but it is illustrative of the widely different approaches being used in this field. The study of the development of the cortex has both benefited from the shifts in paradigm provided by research in other branches of neuroscience and has provided its own light on these branches.

I am grateful for the contributions of the invited speakers and poster presenters, as well as to Yves Christen and Jacqueline Mervaillie, of the Fondation IPSEN, who are solely responsible for an extremely informative and pleasant meeting.

Boston, Massachusetts, U.S.A. Albert M. Galaburda
Summer, 1997

Contents

List of Contributors

Alder, J.
The Rockefeller University, New York, NY 10021, USA

Caviness, Jr., V.S.
Department of Neurology, Massachusetts General Hospital, 25 Fruit Street, Boston, MA 02114, USA

Dambska, M.
Medical Research Center, Polish Academy of Science, Warsaw, Poland

Davidson, N.
The Rockefeller University, New York, NY 10021, and Caltech, Pasadena, CA 91125, USA

Dehay, C.
INSERM Unité 371, Cerveau et Vision, 18 avenue du Doyen Lépine, 69675 Bron Cedex, France

Evrard, P.
Laboratoire de Neurologie du Développement, INSERM CRI 96-03 and Service de Neurologie Pédiatrique, Faculté de Médecine Xavier-Bichat (Université de Paris VII), Hôpital Robert-Debré, 48 Boulevard Sérurier, 75019 Paris, France

Finlay, B.L.
Uris Hall, Cornell University, Ithaca, NY 14853, USA

Galaburda, A.M.
Beth Israel Deaconess Medical Center, East Campus, 330 Brookline Avenue, Boston, MA 02215, USA

Gleeson, J.G.
Department of Neurology, Children's Hospital, Boston, Fegan 11, 300 Longwood Ave, Boston, MA 02115 and Division of Neurogenetics, Department of Neurology, Beth Israel Deaconess Medical Center/Harvard Medical School, Harvard Institute of Medicine, 77 Avenue Louis Pasteur, Boston, MA 02115, USA

Gressens, P.
Laboratoire de Neurologie du Développement, INSERM CRI 96-03 and Service de Neurologie Pédiatrique, Faculté de Médecine Xavier-Bichat (Université de Paris VII), Hôpital Robert-Debré, 48 Boulevard Sérurier, 75019 Paris, France

Hatten, M.E.
The Rockefeller University, New York, NY 10021, USA

Heintz, N.
The Rockefeller University, New York, NY 10021, USA

Hofer, M.
The Rockefeller University, New York, NY 10021, USA

Innocenti, G.M.
Institut de Biologie Cellulaire et de Morphologie, 9 rue du Bugnon, 1005 Lausanne, Switzerland

Kennedy, H.
INSERM Unité 371, Cerveau et Vision, 18 avenue du Doyen Lépine, 69675 Bron Cedex, France

Kida, E.
Department of Pathological Neurobiology, NYS Institute for Basic Research in Developmental Disabilities, 1050 Forest Hill Road, Staten Island, NY 10314, USA and Medical Research Center, Polish Academy of Science, Warsaw, Poland

Kingsbury, M.A.
Institut de Biologie Cellulaire et de Morphologie, 9 rue du Bugnon, 1005 Lausanne, Switzerland

Kofuji, P.
The Rockefeller University, New York, NY 10021, USA

Kuchna, I.
Department of Pathological Neurobiology, NYS Institute for Basic Research in Developmental Disabilities, 1050 Forest Hill Road, Staten Island, NY 10314, USA and Medical Research Center, Polish Academy of Science, Warsaw, Poland

Lester, H.
The Rockefeller University, New York, NY 10021, USA

Marret, S.
Laboratoire de Neurologie du Développement, INSERM CRI 96-03 and Service de Neurologie Pédiatrique, Faculté de Médecine Xavier-Bichat (Université de Paris VII), Hôpital Robert-Debré, 48 Boulevard Sérurier, 75019 Paris, France

Millen, K.
The Rockefeller University, New York, NY 10021, USA

Millonig, J.
The Rockefeller University, New York, NY 10021, USA

Niederer, J.K.
Developmental Neuroscience Group, Cornell University, Ithaca, NY 14853, USA

Nowakowski, R.S.
Department of Neuroscience and Cell Biology, UMDNJ-Robert Wood Johnson Medical School, Piscataway, NJ 08854, USA

Rakic, P.
Section of Neurobiology, Yale University School of Medicine, New Haven, Connecticut 06510, USA

Ross, E.
The Rockefeller University, New York, NY 10021, USA

Soriano, E.
Department of Animal and Plant Cell Biology, Faculty of Biology, University of Barcelona, Diagonal 645, 08028 Barcelona, Spain

Sotelo, C.
INSERM U.106, Hôpital de la Salpétrière, 47 bld de l'Hôpital, 75651 Paris Cédex 13, France

Takahashi, T.
Department of Neurology, Massachusetts General Hospital, 25 Fruit Street, Boston, MA 02114, USA and Department of Pediatrics, Keio University School of Medicine, Tokyo 160, Japan

Tettoni, L.
Institut de Biologie Cellulaire et de Morphologie, 9 rue du Bugnon, 1005 Lausanne, Switzerland

Walsh, C.A.
Division of Neurogenetics, Department of Neurology, Beth Israel Deaconess Medical Center/Harvard Medical School, Harvard Institute of Medicine, 77 Avenue Louis Pasteur, Boston, MA 02115, USA

Wierza-Bobrowicz, T.
Institute of Psychiatry and Neurology, Warsaw, Poland

Wisniewski, K.E.
Department of Pathological Neurobiology, NYS Institute for Basic Research in Developmental Disabilities, 1050 Forest Hill Road, Staten Island, NY 10314, USA

Zheng, C.
The Rockefeller University, New York, NY 10021, USA

Cell Proliferation in Cortical Development

V. S. Caviness, Jr., T. Takahashi and R. S. Nowakowski

Abstract

Neocortical neurons arise in a pseudostratified epithelium (PVE) which forms the lining of the embryonic cerebral ventricular cavity. At any given point on the PVE, the order of formation of the principal projection neuron classes is invariant, beginning with the polymorphic neurons of layer VI, followed by the large pyramids of layer V, the granular neurons of layer IV and finally the medium and small pyramids of layers III and II. This sequence is initiated and propagated across the PVE rostrolaterally to caudomedially as a "cytogenetic gradient." Experiments in mouse have determined that the neuronogenetic process requires 11 cycles over a six-day period. The mechanisms of regulation of the duration of the G1 phase of the cell cycle (T_{G1}) in the PVE in mouse are also the mechanisms that determine the pattern of progression of the duration of the cell cycle (T_C) and, therefore, the number of integer cell cycles that constitute the neuronogenetic interval. Because T_{G1} is regulatory for the number of cell cycles, it is also one of the principal regulatory points for the neuronal output of the PVE. Moreover, for any position on the PVE, there is a tight correlation between the class of neuron being formed at any moment and the value of T_{G1}, and the T_{G1} gradient corresponds to the cytogenetic gradient. Experiments with murine cerebral wall explants *in vitro* indicate that the gradient is driven by a "generator" acting at the conjunction of neocortical and striatal PVE at the lateral ventricular angle. The mechanism of gradient propagation is independent of the generator and may be mediated by gap junctions. The slope of the gradient appears to be modulated by β-FGF. We propose that mechanisms regulatory to T_{G1} are also regulatory to the determination of neuronal class. We propose further that the gradient in T_{G1} is the mechanism of positional encoding of a neocortical architectonic protomap within the PVE. Both of these histogenetic regulatory functions of T_{G1} would require that there is a systematic and coordinate alteration in the gene transcriptional profile with each step advance in T_{G1} with progression through the successive integer cell cycles of the neuronogenetic interval.

Introduction

Neuronogenesis is the process of cell proliferation that results in the production of the building blocks, i.e., the neurons, of the central nervous system. The cell

A. M. Galaburda / Y. Christen (Eds.)
Normal and Abnormal Development of Cortex
© Springer-Verlag Berlin Heidelberg New York 1997

proliferative activity that gives rise to neocortical neurons occurs in a pseudostratified epithelium (PVE) which form the ventricular margin of the embryonic cerebral vesicle and is approximately co-extensive with the architectonically defined ventricular zone (VZ; reviewed in Takahashi et al. 1993, 1995; Fig. 1). The area of the neocortical PVE is extensive, forming the lining of the lateral and dorsal convexity of the cerebral ventricular system (Bayer and Altman, 1991). The neocortical PVE is continuous at the lateral ventricular angle with the PVE of the striatum and medially with the PVE of the paleocortex and archicortex.

Postmitotic young neurons exit the epithelium to migrate across the subventricular and intermediate zones of the cerebral wall and take up positions within the neocortex emerging at the cerebral surface (Sidman and Rakic 1973; Fig. 1). Neurons are formed and migrate in an "inside-to-outside" sequence with respect to their eventual positions in the cortex. Thus, the earliest neurons to arise will form the deepest cortical layer whereas successively later arising neurons will form the successively more superficial layers. Once in position within the cortex, the young neurons become linked by synaptic junctions into the neural circuits that will support the operation of the overall neural systems of the forebrain.

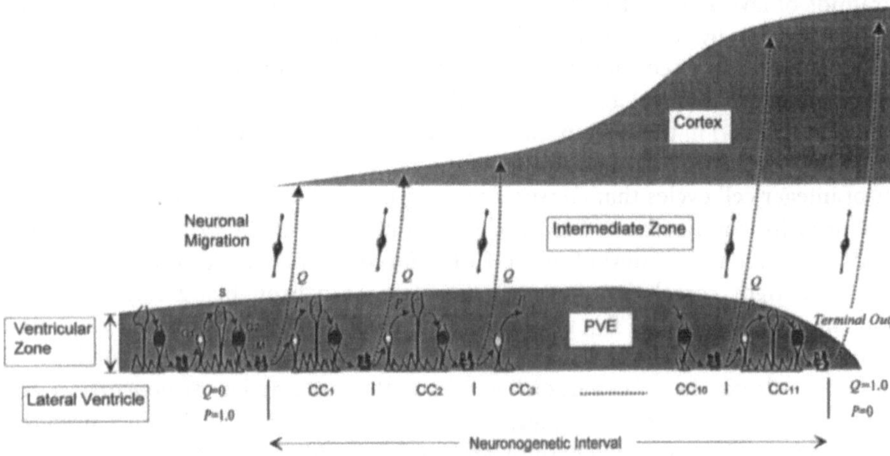

Fig. 1. Schematic representation of the developmental changes in the neocortical cerebral wall during the neuronogenetic interval. The pseudostratified ventricular epithelium (PVE) is approximately coextensive with the ventricular zone lying at the margin of the lateral ventricle. Young postmitotic neurons migrate across the intermediate zone of the intervening cerebral wall (broken arrows) to find their final positions in the developing cortex at the surface of the hemisphere (neuronal migration). The proliferative cells of the PVE undergo interkinetic nuclear migration (curved arrows) in the course of the cell cycle which begins with G1 phase and progresses through S and G2 and is completed with M phase. Postmitotic cells either exit the cell cycle to initiate migrations (Q fraction, *Q*) or elongate through G1 phase to re-enter S phase (P fraction, *P*). Prior to the onset of neuronogenesis, by definition, the P fraction is 1.0 and the Q fraction is 0. Q becomes greater than 0 with the onset of neuronogenesis and eventually reaches 1.0 at the end of the neuronogenetic interval, which corresponds to 11 integer cell cycles (CC_1-CC_{11}) in mouse. P, reciprocally, follows a path from 1.0 to 0. The PVE gradually increases in height through much of the neuronogenetic interval. Over the final cycles, however, it rapidly regresses and ceases to exist with the termination of CC_{11}. At the completion of CC_{11} (the last cell cycle of neuronogenesis in mouse) the P fraction cells each will divide into two daughter cells, all of which exit the VZ as Q (*terminal output*; adapted from Takahashi et al. 1996 b)

Ultimately a substantial proportion, perhaps as much as 50 %, of intracortical neurons will be eliminated by histogenetic cell death (Ferrar et al. 1992; Finlay and Pallas 1989; Takahashi et al. 1996c).

Three cardinal achievements of the overall neocortical histogenetic process have been ascribed to the proliferative process: 1) formation of neurons in numbers appropriate to the requirements of neocortical histogenesis (Takahashi et al. 1996 b), 2) the determination of neuronal class (McConnell 1991) and 3) the specification of the neocortical architectonic fields that partition the neocortex into neural systems-related subdivisions (Rakic 1988). The first of these is an established and now well-characterized role of the PVE (Caviness et al. 1995; Takahashi et al. 1996 b). Experimental support for the contribution of the PVE to the other two histogenetic achievements remains circumstantial at present.

We review here the workings of the proliferative process as it operates within the PVE to produce neurons in numbers appropriate to neocortical histogenesis. We also propose a theoretical framework within which mechanisms fundamental to the proliferative process might also be deterministic to specification of neuronal class and the neocortical architectonic map. Specifically, we will highlight mechanisms of regulation of a single proliferative parameter, the duration of the G1 phase of the cell cycle (T_{G1}), as the theoretical central regulatory mechanism underlying the numerical pattern of neuronal production, as well as the specification of neuronal class and the neocortical architectonic map.

The Proliferative Process in the PVE

Cellular Organization and Interkinetic Nuclear Migration

The PVE, as a pseudostratified epithelium, is homogeneous throughout in appearance and cellular organization. It is constituted of radially elongate cells attached to each other at the ventricular surface by adherens junctions and linked by gap junctions (Caviness et al. 1995; Dermietzel and Spray 1993; Hinds and Ruffett 1971; Lo Turco and Kriegstein 1991; Stensaas and Stensaas 1968). As a consequence of interkinetic nuclear migration, the radial position of cell soma and nucleus is systematically correlated with its phase in the cell cycle (Fujita 1960; Sauer 1936; Fig. 1). Thus, S phase occurs with the cell nucleus in the outer half of the VZ. The nucleus descends as the cell goes through G2 and prophase. Mitosis occurs with the soma and nucleus at the ventricular margin.

Prior to the onset of neuronogenesis, all postmitotic cells return to the proliferative pool and growth of the PVE is exponential (Takahashi et al. 1996 b). With the onset of the neuronogenetic interval, that is, the interval through which postmitotic neurons arise from the PVE, a postmitotic daughter cell may have only one of two possible proliferative fates. The cell may either leave the cycle as a permanently postmitotic young neuron (Q or quiescent fate) or it may re-enter the proliferative pool (P or proliferative fate; Takahashi et al. 1994, 1996 b, c; Fig. 1). If the daughter cell elects Q, further progression through the cell cycle is arrested, and the now postmitotic daughter cell will ascend and exit the VZ. If the daughter cell elects P, its nucleus in G1 phase will ascend only across the width of the PVE before it re-enters S phase. The Q fraction of the PVE is the fraction of postmitotic PVE daughter cells that elects the Q fate. The Q fraction will progress during the neuronogenetic period from 0 to 1.0 P is complementary to Q (P=1-Q) where cell death is considered to be negligible.

Neuronogenetic Interval

The onset and termination of the neuronogenetic interval are discrete. Onset corresponds to the moment when the Q fraction, or simply "Q," becomes non-zero and the P fraction, or simply "P," becomes less than 1.0, whereas termination corresponds to the moment that Q becomes 1.0 and P becomes 0 (Takahashi et al. 1996 b). Over the intervening neuronogenetic interval the progeny of the founder population will execute a series of integer cell cycles (Takahashi et al. 1995). The profile of neuronal classes which emerges with each integer cycle is unique and changes continuously as the series of cycles advances. Thus, in any given zone of the PVE, the principal projection neuron classes of the separate neocortical layers are formed in an identical deep to superficial ("inside out") sequence (Caviness 1982; Rakic 1974; Sidman and Rakic 1973) as the successive integer cycles are executed. The cell class succession is as follows: the polymorphic neurons of layer VI, the large pyramidal neurons of layer V, the granular neurons of layer IV and the medium and small pyramidal neurons of layers III and II (Caviness 1982). Cellular proliferative behavior (specifically the progression of the values for T_{G1} and Q), as the succession of neuron classes is formed in each zone of the PVE, is invariant throughout the PVE. Interneurons, cells which are similar to each other in their morphological features and are distributed more or less uniformly to all cortical layers, are formed concurrently with the projection neurons throughout the neuronogenetic interval (Lavdas et al. 1996; Miller, 1985; Parnavelas et al. 1991; Peters and Jones 1984). It is uncertain if the contribution of interneurons to the Q fraction varies in the course of integer cycle progression.

Cytogenetic Gradient

Although the PVE is homogeneous in its cytological organization and gives rise throughout to the same cell class succession, the phasing of cytogenetic events is systematically shifted gradient-wise across the epithelial sheet (Fig. 2). This gradient is established because the outset of neuronogenesis occurs first in the most rostrolateral region and then progresses towards the caudomedial pole of the PVE sheet. Thus, the polarity of the gradient is rostrolateral to caudomedial. Historically the gradient has been designated the "cytogenetic gradient" because it has been defined in terms of the sequence of formation of neuronal classes (or neocortical layer; Bayer and Altman 1991; Bisconte and Marty 1975; Fernandez and Bravo 1974; Hicks and D'Amato 1968; Smart and McSherry 1982; Smart and

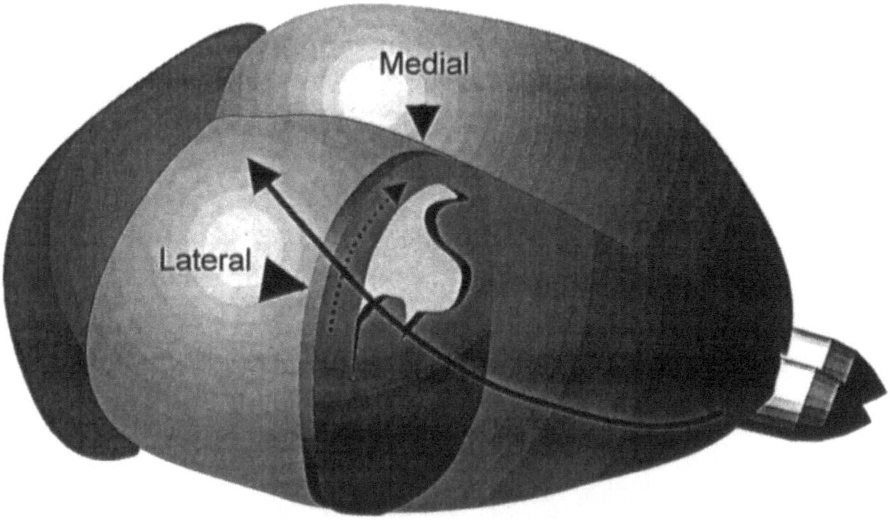

Fig. 2. The neocortical cytogenetic gradient. The first neocortical neurons that undergo terminal divisions do so in the rostrolateral PVE. Cytogenesis then propagates from rostrolateral to caudomedial across the full expanse of the PVE (thick curved arrow). At any given coronal plane, there will be a lateral to medial gradient (insert, broken arrow)

Smart 1982). The steepness of the cytogenetic gradient varies over the course of the neuronogenetic interval. In mouse, for example, formation of the first permanent neuronal population, the polymorphic cells destined for layer VI, occurs rostrolaterally in the PVE late on E10 (Smart and McSherry 1982: Smart and Smart 1982; Wood et al. 1992). In progressively more caudomedial regions of the PVE, formation of the polymorphic neurons of layer VI is delayed progressively later until early E11 (Caviness 1982). The slope of the cytogenetic gradient in mouse rises slowly through E12 as the deeper neurons of layer VI are being formed throughout the PVE (Smart and McSherry 1982; Smart and Smart 1982; Fig. 3 A). At any given time, those formed rostrolaterally are destined for slightly more superficial cortical positions than those formed progressively caudomedially. The slope rises much more rapidly between E12 and E14, in that granule cells destined for layer IV and medium neurons destined for inner layer III, respectively, are being formed rostrally and laterally while, simultaneously, polymorphic neurons destined for outer layer VI and large pyramidal neurons of layer V are being formed caudomedially (Fig. 3 B). The slope of the cytogenetic gradient then flattens again after E14. It remains nearly flat through the end of the neuronogenetic interval, which occurs late on E16 to early on E17. In this final "flat" phase of the neuronogenetic interval, medium and small pyramidal neurons destined for the most superficial level of layer III and for layer II, respectively, are being formed throughout the PVE (Fig. 3 C).

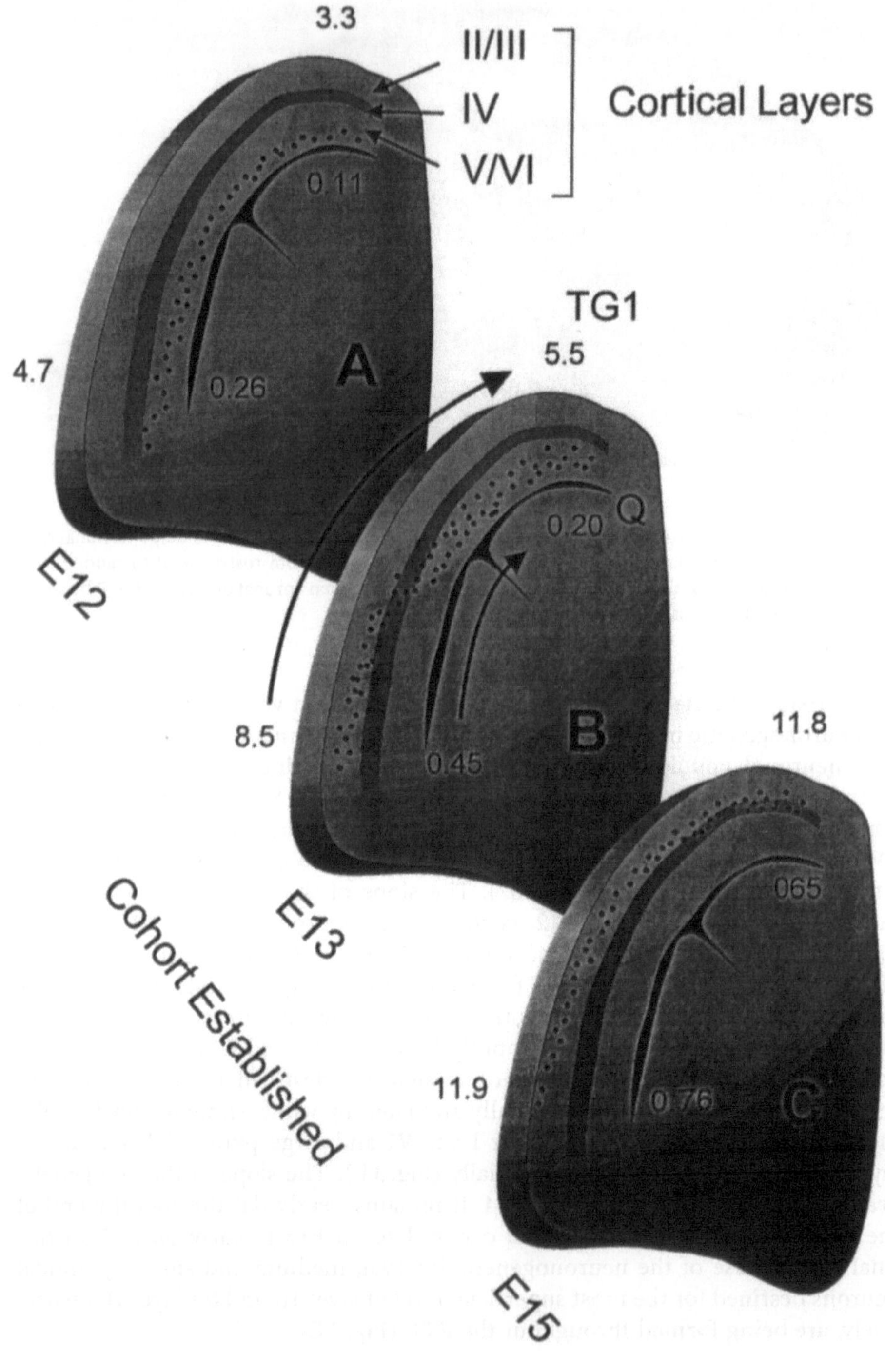
3.3
II/III
IV
V/VI
Cortical Layers
0.11
TG1
4.7
A
0.26
5.5
Q
0.20
E12
8.5
B
0.45
11.8
065
Cohort Established
E13
11.9
0.76
C
E15

◀ **Fig. 3.** The cytogenetic gradient in relation to PVE gradients in T_{G1} (values above cortex in each plane, A–C) and in Q (values below cortex in each plane, A–C). The positions of neurons "born" on E12 (A), E13 (B), and E15 (C) are plotted (dots) with respect to neocortical layers (II–VI). The cytogenetic gradient, thus represented, is shallow on E12, steep on E13 and shallow and E15. The gradient in T_{G1} is similarly steepest on E13, on which date there is a 3.0 hr difference (8.5 and 5.5 hr) in values of T_{G1} laterally and medially. On E12, by contrast, the lateral to medial differential in T_{G1} is only 1.4 hr, whereas on E15, T_{G1} is equal laterally and medially. The lateral to medial gradient in Q, by contrast to that in T_{G1}, is relatively great at each date

The Proliferative Process

We have characterized the proliferative process quantitatively in mouse through a series of experimental analyses based either upon cumulative labeling with bromodeoxyuridine (BUdR, cell cycle parameters, Tc and T_{G1}; Takahashi et al. 1992, 1993, 1995) or a protocol involving successive exposure to ^{3}H-thymidine then BUdR (Q and P fractions; Takahashi et al. 1994, 1996 b). To define the proliferative correlates of the cytogenetic gradient, the analyses have been undertaken in both medial and lateral zones of the neocortical PVE, zones that are widely separated from each other along the cytogenetic gradient (Miyama et al., 1997; Fig. 3). There are three principal findings.

Uniformity of the proliferative process throughout the PVE

Although neuronogenesis is initiated and completed earlier laterally than medially, respecting the cytogenetic gradient (Fig. 3), the overall duration of the neuronogenetic interval laterally and medially is identical at six days. Both laterally and medially the growth fraction (GF, the fraction of cells of the PVE which is proliferating) is held constant near 1.0, and in both regions of the PVE the duration of the cell cycle (T_C) advances from approximately 10 to approximately 20 hr. The path of progression of T_C corresponds both laterally and medially to a division of the overall neuronogenetic interval into 11 integer cell cycles. That is, each region of the PVE operates through an identical proliferative interval (= six days) representing an identical number of integer cell cycles (= 11 cycles). For each region of the PVE the range and pattern of progression from minimum to maximum values of proliferative parameters (i.e., T_{G1} and P and Q fractions) are identical despite the fact that phasing of such progression will be quite different in regions widely separated across the cytogenetic gradient (Fig. 4).

T_{G1} is the only developmentally regulated phase of the cell cycle

Laterally and medially T_{G1} advances through an identical range of approximately 4 to 12 hr (Fig. 4), accounting for the near doubling of T_C over the neuronogenetic interval. The durations of neither S phase (T_S) nor G2+M phases (T_{G2+M}) vary

Fig. 4. The length of the G1 phase (A) and Q (B) are plotted as a function of elapsed integer cell-cycles in medial (×) and lateral (•) PVE. The progression of values, thus normalized with respect to integer cycles, is closely correspondent. The steepest part to the slope for T_{G1} corresponds to the ascent of Q through 0.5, where the number of cells leaving the cycle is the same as the number of those remaining. T_{G1} reaches asymptote as Q reaches its terminal value of 1.0 at the completion of neuronogenesis (Adapted from Miyama et al. 1997). ▶

systematically through the neuronogenetic interval. We emphasize that the number of integer cycles which constitutes the neuronogenetic interval is uniquely determined by the advance of T_{G1} through its range of minimum to maximum values.

Q (and P) varies identically with T_{G1} across the PVE

Q ascends (with descent of P which is 1-Q) across the neuronogenetic interval. The pattern of ascent of Q from 0 to 1.0 (and complementary descent of P from 1.0 to 0) with respect to T_{G1} (and, therefore, also to integer cycle) is virtually identical across the medial to lateral expanse of the PVE. The paths of ascent of Q are non-linear, rising slowly at first but more rapidly toward the end of the neuronogenetic interval (Fig. 4 B). The steady state value of Q = P = 0.5 is not reached in mouse until the eighth of the 11 integer cycles, corresponding to an elapse of approximately 80% of the neuronogenetic interval. T_{G1} like Q rises gradually at first but ascends rapidly to asymptote while Q is still increasing and in so doing diverges from the Q (Fig. 4 A). The divergence of the patterns of progression of Q and T_{G1} with respect to integer cycle is maximum around the point where Q=0.5 both medially and laterally in the PVE.

Proliferative Mechanisms and Operation of the Cytogenetic Gradient

T_{G1} is distributed as a gradient across the PVE, and the slope of this gradient changes systematically through the course of the neuronogenetic interval (Miyama et al., manuscript in preparation; Fig. 5). We infer from these observations that cell-cell communication critically modulates the cell internal mechanisms regulatory to T_{G1}. Two sets of observations based upon studies conducted *in vitro* with E13 explants of the cerebral wall have implications for cell external mechanisms of T_{G1} modulation. E13 corresponds to the sixth integer cycle medially but to the eighth integer cycle laterally. In both sets of experiments, the findings are in the form of BUdR labeling indices (LI), following five hours of exposure to the marker. At E13, the LI in the PVE is distributed as a lateral to medial rising gradient because T_{G1} is shorter medially than laterally (Fig. 5).

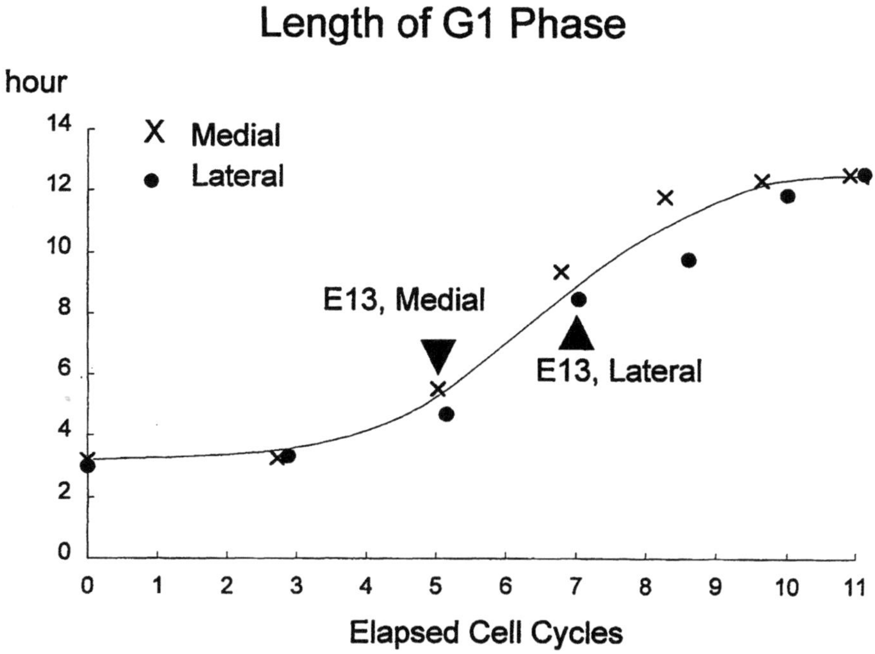
Length of G1 Phase
hour
14
12
10
8
6
4
2
0
X Medial
● Lateral
E13, Medial
E13, Lateral
0
1
2
3
4
5
6
7
8
9
10
11
Elapsed Cell Cycles

a

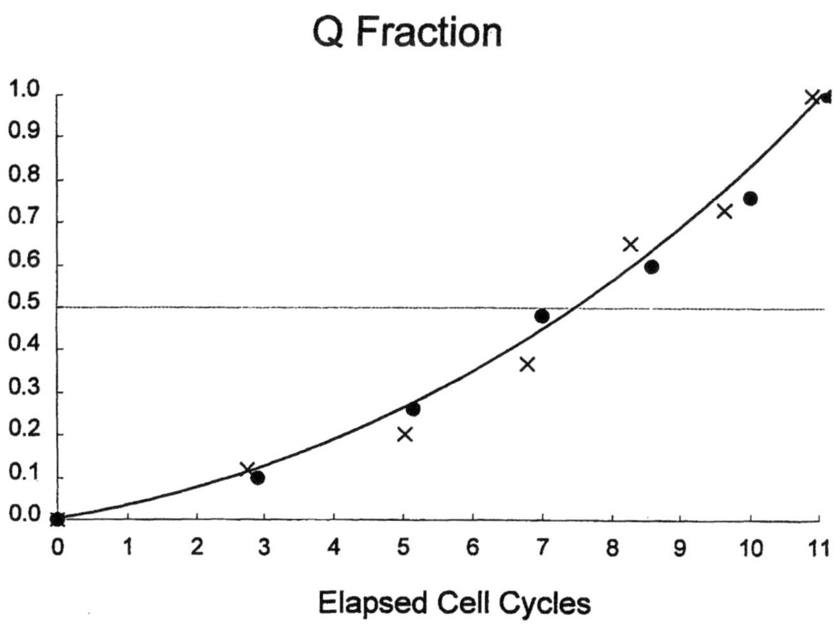
Q Fraction
1.0
0.9
0.8
0.7
0.6
0.5
0.4
0.3
0.2
0.1
0.0
0
1
2
3
4
5
6
7
8
9
10
11
Elapsed Cell Cycles

b

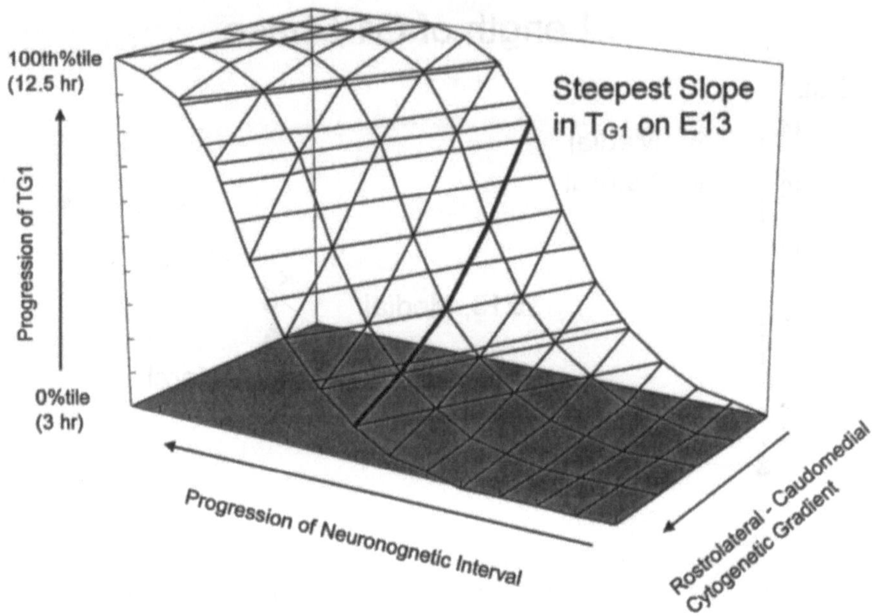

Fig. 5. Three dimensional plot of the relationship between the progression of T_{G1}, the progression of the neuronogenetic interval and the lateral to medial cytogenetic gradient. The range of T_{G1}, represented on the ordinate as percentile excursion, is 3–12.5 hr. The T_{G1} slope, like that of the cytogenetic gradient, is steepest on E13 (thick line)

A Gradient Generator

The LI gradient is flattened by interruption of continuity of striatal and neopallial PVE at the lateral ventricular angle (Miyama et al. 1995, and unpublished observations). Specifically, the LI laterally becomes increased, approaching that medially (Fig. 6). Although we have not yet measured T_{G1} in this system, we infer from theoretical considerations, to be considered subsequently, that the effect of the PVE interruption at the striatal neopallial boundary is to reduce the length of T_{G1}. The effect is seen increasingly laterally where T_{G1} is most advanced across the range of minimum to maximum values. It is not seen medially where T_{G1} is still near its minimal value.

From this analysis we infer that some property of the system which is dependent upon the integrity of the continuity of striatal and neopallial PVE (or possibly some other adjacent structure) acts locally to prolong T_{G1} in the lateral PVE. Without implications for either its structural or physiological nature, we refer to this property as the "T_{G1} gradient generator." The gradient in T_{G1} is tonically maintained and lengthened by this generator throughout the neuronogenetic interval. The gradient in T_{G1} is elastic. If the neocortical PVE is separated from the generator, there is recoil of the values of T_{G1} and this recoil is graded along the lateral to medial axis of the PVE such that the longer the T_{G1} at PVE interrup-

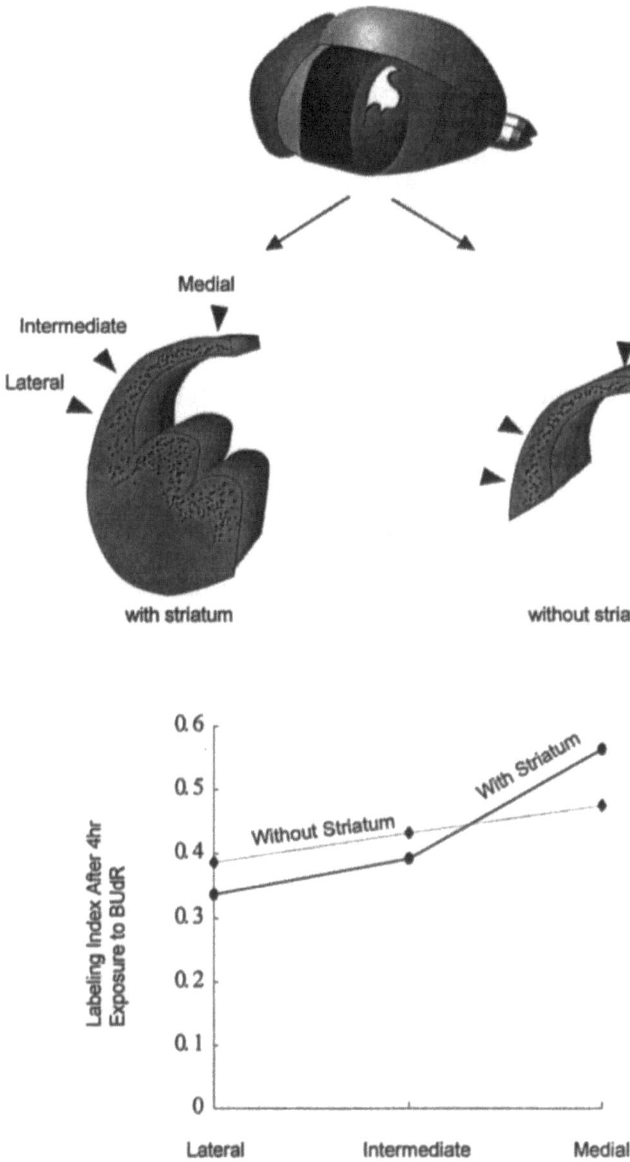

Fig. 6. Consequences of transection of the cerebral wall at the lateral ventricular angle upon the gradient in T_{G1} in cerebral wall explants. Explants, removed by trocar from the convexity of the cerebral wall of E13 mouse embryos (insert with "white ventricular outline"), are cultured in collagen gel in minimal medium. In an experimental but not in a control series, the cerebral wall is transected at the lateral ventricular angle, as illustrated, at the time the explant is removed. The S-phase marker bromodeoxyuridine (BUdR) is added after 5 hr in culture, and exposure to the tracer is continued for an additional 4 hr. The BUdR labeling index (labeled cells as proportion of total cells) is determined in sectors of PVE located medially, at an intermediate position and laterally with respect to the coronal plane of section (arrowheads). The transection is associated with an upward recoil in the slope of the lateral to medial gradient in the BUdR labeling index

tion, the greater the degree of recoil to basal levels of T_{G1}. The coherency of propagation of the effect of PVE interruption, occurring over hundreds of microns at a latency of hours or less, requires, we suggest, a high precision mechanism of cell-cell communication. As hypothesis we propose that such a mechanism could be gap junction-mediated. Gap junctions have been invoked as involved in sustaining gradients in invertebrate developmental fields (Bennett et al. 1991; Goodwin and Cohen 1969; Wolpert 1969, 1978). As noted earlier, they are known to link cells of the PVE (Dermietzel and Spray 1993; Lo Turco and Kriegstein 1991).

Modulation of the Gradient

β-FGF added to the culture medium for an interval less than T_C, like interruption of continuity of striatal and neopallial PVE, alters the LI profile in the PVE of E13 mouse in a way that suggests a decrease in T_{G1} (Fig. 7). The effect is seen only laterally, where T_{G1} is most advanced through its normal minimal to maximal range. It is not seen medially, where T_{G1} is still near its minimal value. If this interpretation is correct, the implication is that β-FGF modulates the gradient in T_{G1} at E13. The greater the prolongation of T_{G1} above the minimal level, the greater will be the effect of β-FGF. We infer additionally that β-FGF, presumably among a variety of diffusible cell external substances, acts to set the sensitivity of gap junction-mediated mechanisms of T_{G1} gradient propagation. That is, we sug-

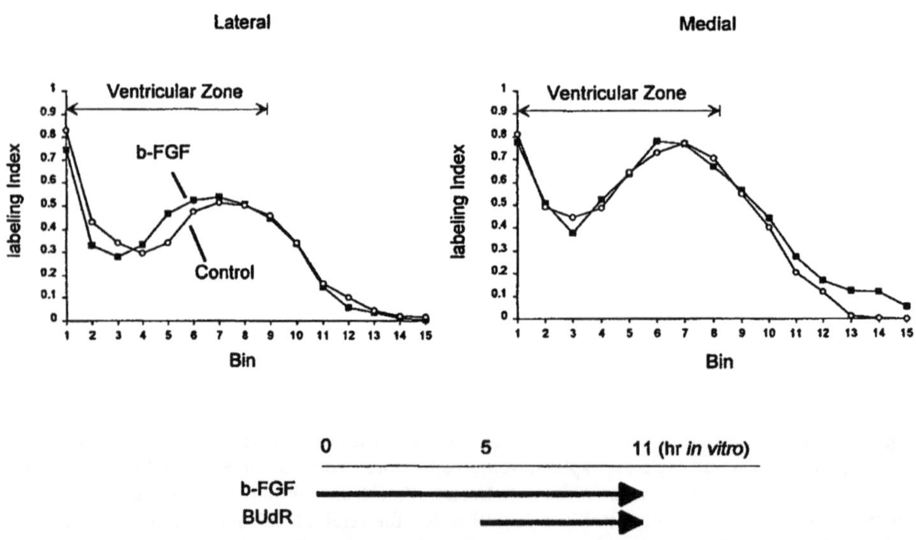

Fig. 7. The effect of β-FGF upon T_{G1} in cerebral wall explants. Explants of the E13 cerebral are established as described in the legend to Figure 6. β-FGF is added to an experimental but not to a control series of explants from the time of explantation through a total of 11 hr *in vitro*. BUdR is added during the final 6 hr *in vitro*. With β-FGF there is an augmentation of the BUdR labeling index of the ventricular zone in the lateral but not in the medial region. This observation is consistent with a decrease in T_{G1} with β-FGF laterally but not medially (Goto et al. 1996)

gest that diffusible substances may act to modulate the slope of the gradient but are not the mechanism of its propagation.

We propose a four-component model for the overall mechanism that establishes and maintains the T_{G1} gradient: 1) cell internal mechanisms that determine the (species characteristic) potential range of T_{G1}, 2) a gradient-generating mechanism acting tonically at the striatal-neopallial junctional region which drives T_{G1} from the minimum to maximum values of this range, 3) a mechanism of cell-cell communication that propagates the effect of the gradient generator (gap junctions as candidate mechanisms) and 4) a resultant effect of diffusible, positively and negatively T_{G1} modulatory, cell external substances interactive with cell surface receptors, which act to shape the contour of the gradient.

Within this framework, T_{G1} would be constrained between maximum and minimum values whereas the specific shape of the gradient over the neuronogenetic interval would be determined by the resultant effect of the concentrations of cell external modulators. One would predict, for example, that the positive modulatory effect of cell external factors would be maximum in regions of the PVE where cells of layers V-III are being formed and the slope of the T_{G1} gradient is maximum (Takahashi et al. 1996a). The valence of modulators might be regulated by either or both their concentrations or the density of their receptors on the cell surface.

The Proliferative Process and Histogenesis

Neuron Production

Neuronal output from a PVE founder population of specified size is determined by three principal parameters: 1) the growth fraction, 2) the number of integer cycles which comprises the neuronogenetic interval and 3) the progression of Q and P with respect to integer cycle (Caviness et al. 1995; Takahashi et al. 1996b). Formulated simply, the output of a founder population of growth fraction 1.0 will be the sum of the outputs of the full series of integer cycles. The number of cycles will be determined by the duration of the neuronogenetic interval, the initial value of T_C and, importantly for our general emphasis upon its regulatory role in histogenesis, the progression of T_{G1}. These parameters, as noted in the foregoing sections, have been determined across the neurogenetic interval in mouse. The output for each cycle will be determined by the size of the proliferative pool at the outset of the respective cycle and Q for that cycle.

Growth of the PVE and Neuronal Output

Expansion of the PVE and neuronal output are initially slow but then accelerate (Fig. 8). The PVE reaches its maximum size at $Q = P = 0.5$ after an expansion of approximately 55-fold with respect to the founder population (Takahashi et al.

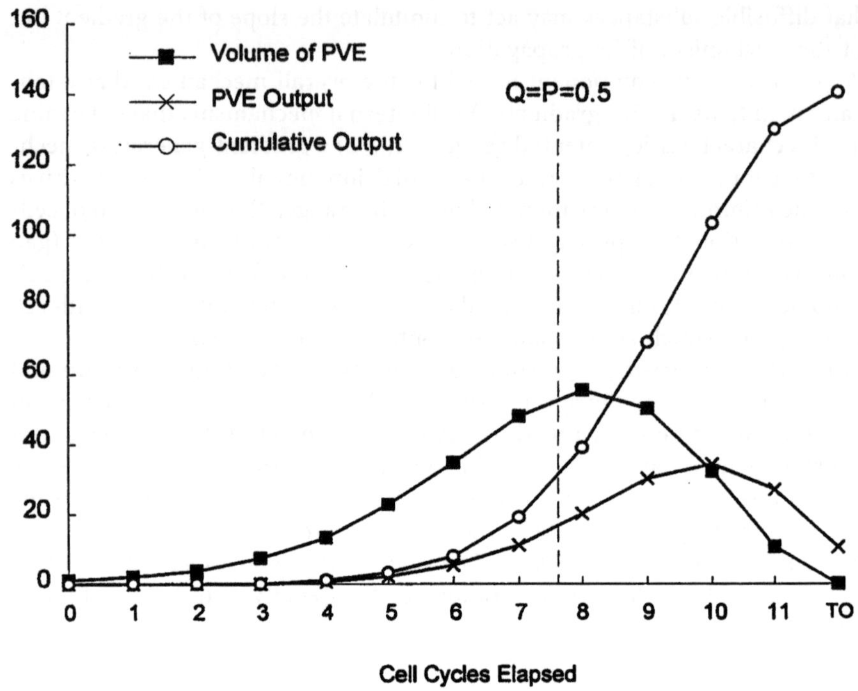

Fig. 8. Expansion and regression of a founder PVE population and cell output over the course of the neuronogenetic interval. The volume of the PVE (■), cell output from a single cell cycle (×) and the cumulative cell output (o) are plotted with respect to the elapsed cell cycles (Takahashi et al. 1996). At the beginning of the neuronogenetic interval, that is, at the point where 0 cell cycles have elapsed (i. e., the beginning of CC_1) at 9:00 am on E11, the PVE volume is set at the arbitrary unit value of 1.0, and cell output at this point is by definition 0. The vertical dashed line where Q and P reach 0.5 separates the neuronogenetic interval. To the left of the vertical dashed line, $Q<0.5$ and the PVE is expanding; to the right of the vertical dashed line, $Q>0.5$ and the PVE is regressing, through a massive increase in its output. Note that the PVE size reaches its maximum value at this point and cell output/cycle is maximum beyond this point. The P fraction cells of the last cell cycle (eleventh in mouse) will divide to produce two daughter cells, both of which will exit the VZ as the terminal output (Q=1.0, terminal output or TO on the abscissa). (Adapted from Takahashi et al. 1996 b)

1996b). Subsequently the PVE regresses and ceases to exist with the termination of neuronogenesis. As Q advances from 0.5–1.0 and the termination of neuronogenesis approaches, there is a progressive and rapid acceleration of neuronal output. Overall the average founder cell (a cell proliferating at the time of initiation of neuronogenesis on late E10 to early E11, depending on location in the gradient, in mouse) will give rise to approximately 150 cortical neurons. The impact of cell death upon growth of the PVE and neuronal output is unknown but is probably small. The recent analysis of Blaschke et al. (1996), which holds that cell death in the murine PVE is equal to or greater than 50 % throughout the neuronogenetic interval, would preclude all growth of PVE and would preclude a crescendo of neuronal output late in the neuronogenetic interval, both of which are incontrovertible realities of the PVE proliferative process. The ISEL + method used by these investigators, though in all probability very sensitive to cell death, has not been shown to be diagnostic for cell death and is apparently not specific to cell death. More certain, however, is the impact of cell death upon the postmigratory population of the neocortex. This appears to be substantial, perhaps as high as 15–20 % in infragranular (Heumann et al. 1978; Takahashi et al. 1996 c) and 50 % in supragranular cortical layers (Ferrar et al. 1992; Finlay and Pallas 1989).

Cell Class and the Progression of T_{G1}

T_{G1} and the Cytogenetic Gradient

Our investigations indicate that the proliferative correlate within the PVE of the cytogenetic gradient is a gradient in T_{G1}. The evidence is that the slope of the cytogenetic gradient and the gradient in T_{G1} across the PVE are virtually identical (Miyama et al., 1997; Takahashi et al. 1996 a). Thus, through E12, when the slope of the cytogenetic gradient is relatively flat, that of the gradient in T_{G1} is also relatively flat (Fig. 5). That is, there is only slight prolongation of T_{G1} across the medial to lateral axis of the PVE. From E12 through E14 the slopes of both the cytogenetic and T_{G1} gradients ascend steeply. Thus, on E13 cells of layer III are being formed laterally while cells of layer VI are formed medially, and this differential in the cell class sequence is associated with a maximum differential in T_{G1} of 3 hr across the lateral to medial axis of the PVE. After E14, the slopes of both gradients flatten again. Thus, beyond E15 neurons of layers III/II are being formed laterally and medially and the values of T_{G1} have equalized.

T_{G1} and Cell Class

The critical generalization to extract from the closely parallel behavior of these gradients is that formation of each of the succession of principal neuron classes is associated with a cell class-characteristic value of T_{G1} (Miyama et al., 1997; Takahashi et al. 1996 a). That is, for any given value of T_{G1}, independently of when it is reached at a point on the PVE and where that point is located there will be a specific and invariable correlation with the principal projection neuron class that is being formed (Fig. 9). We suggest, as a hypothesis, that cell class is not only coordinate with T_{G1} but that the mechanisms that underlie cell class progression are in part regulated by the normal progression of T_{G1}. Implicit in this hypothesis is that each step advance in T_{G1} carries with it a step change in the transcriptional profile of cells that experience that advance in T_{G1}.

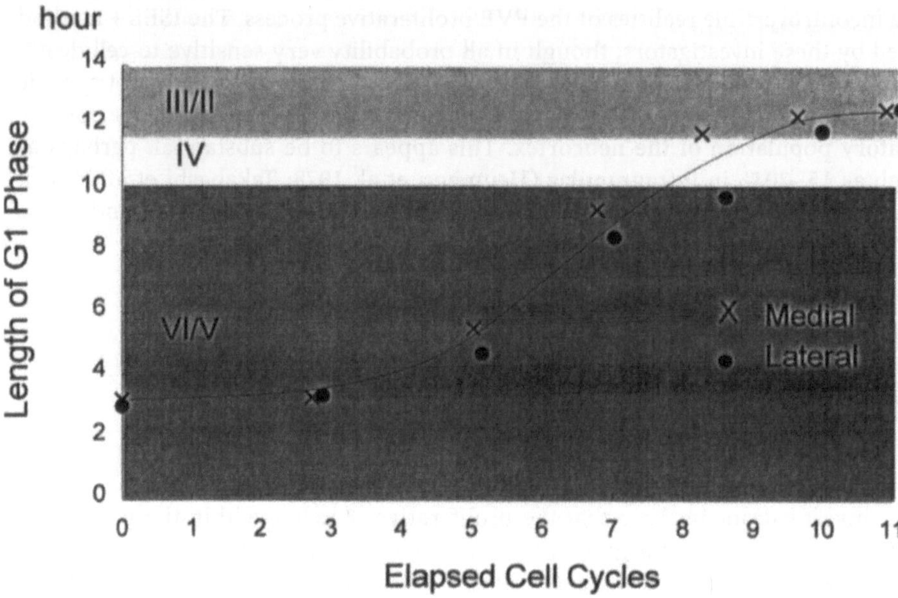

Fig. 9. Correlation of the length of G1 phase and the neocortical layers that are being formed in the PVE. Through the greater part of the ascent of T_{G1}, corresponding to 8 of the 11 integer cycles of the neuronogenetic interval, neurons destined for layers VI and V are formed. Over the remaining ascent of T_{G1}, as the final three integer cycles are completed, neurons destined for layer IV and finally layers III and II are formed. This relationship between T_{G1} and class of neuron being formed is the same in medial (×) and lateral (•) PVE, zones which are widely separated with respect to the cytogenetic gradient

The Neocortical Map and the PVE Gradient in T_{G1}

A Neocortical Protomap within the PVE

Neurons migrate across the cerebral wall to enter the developing cortex in the order in which they undergo their terminal mitoses (Fig. 1). Roughly considered, the neighborhood relationships of cells at their points of origin in the PVE approximate their neighborhood relationships as postmigratory cells within the cortex, although significant intermixing does occur (Austin and Cepko 1990; Misson et al. 1991). The fact of an orderly transposition of the PVE upon the cortex has suggested that the neocortical map is initially instantiated within the proliferative cells of the PVE, that is prior to their terminal divisions and migration to the cortex (Rakic 1988). Although the existence of a "protomap" within the PVE is for the present only conjectural, currently available experimental evidence does establish that the neocortical map becomes instantiated at the level of the cortex at least as early as E15–16 (Erzurumlu and Jhaveri 1992; Molnar and Blakemore 1992; O'Leary et al. 1994). Instantiation of the map is documented specifically with respect to cells of layers VI/V, that is to cells formed as late as E13 or E14 in mouse (Caviness 1982; Takahashi et al. 1996 a). The evidence is that reciprocal, topologically ordered mapping of thalamocortical and corticothalamic projections is established with respect to the postmigratory neurons of layers VI and V (Caviness 1988; Erzurumlu and Jhaveri 1992; Molnar and Blakemore 1992; O'Leary et al. 1994). This is relatively early in the course of cortical histogenesis, that is before the assembly of layers IV and III, which are ultimately to be the principal targets of the thalamic projection. Architectonic features distinctive of each field have already begun to emerge within the pattern of cell and fiber arrangement in layers VI and V as early as E16 in mouse (Caviness 1988).

The T_{G1} Gradient and the Protomap

The mechanisms of mapping the neocortex are linked hypothetically to regulation of T_{G1} by the following considerations. As reviewed in earlier sections of this discussion, the PVE is a bounded proliferative epithelial sheet with rostral to caudal and lateral to medial gradients in T_{G1}. As such the PVE has the properties of a developmental field (Goodwin and Cohen 1969; Meinhardt 1982; Wolpert 1969) where the cellular mechanism of the gradient is a systematic rostral to caudal and lateral to medial shift with time in the oscillatory frequency of the proliferative cycle. The importance of developmental field theory to mapping in neocortical histogenesis is that it introduces a mechanism by which positional values (and therefore mapping coordinates) may be encoded into cells of the PVE (Goodwin and Cohen 1969; Meinhardt 1982; Wolpert 1969). We suggest, as a hypothesis, that the map (protomap: Rakic 1988) is encoded by the gradient in T_{G1} and this encoding is set at a time in the neuronogenetic interval when the potential spatial resolution in the gradient in T_{G1} is maximum. This would correspond to about E13, which is when the range of T_{G1} across the PVE is maximum and when the slope across the axes is steepest (Miyama et al., 1997; Fig. 5). Again, the hypothesis presupposes that each of the differential values of T_{G1} prevailing across the PVE at the time of encoding of positional values carries a unique transcriptional profile that implements the encoding process.

Regulation of T_{G1} in Vertebrate Cells

Our investigations indicate that T_{G1} advances systematically across the neuronogenetic interval and that the cytogenetic gradient corresponds to a proliferative gradient in T_{G1} established rostrolaterally to caudomedially across the full expanse of the PVE. These observations argue that regulation of T_{G1} is fundamental to regulation of the overall neocortical histogenetic process. We have reviewed evidence, both explicit and circumstantial, that regulation of T_{G1} is pivotal to mechanisms that determine neuronal output from the PVE and that determine both the succession of neuronal class and the specification of the architectonic map of the neocortex. Fundamental to these hypotheses are the assumptions that T_{G1} is regulatable by cell external mechanisms, such as influences of growth factors, that regulation of T_{G1} is systematically coordinated with cell cycle exit (progression of Q fraction) and that regulation of T_{G1} is associated with the systematic specification of the cell transcriptional profile (cell class profile). Observations in vertebrate cells *in vitro*, to be reviewed below provide support for all three of these assumptions.

Modulation of T_{G1} and the Cell Cycle

The transition from G1 to S phase (G1/S transition) is ultimately the event that defines the duration of T_{G1} (Murray and Kirschner 1991). The vertebrate cell is driven through G1 to the G1/S transition by a chain of cell internal molecular mechanisms which extends from without to within the nucleus of the cell (Elledge and Harper 1994; Murray and Hunt 1993; Sherr 1993; Sherr and Roberts 1995). An extensive body of evidence in dissociated cell systems indicates that the cell internal mechanisms for regulation of T_{G1}, but of no other phase of the cell cycle, may be modulated by influences external to the cell (Coats et al. 1996; Massague and Polyak 1995). Thus, in theory, it is the responsivity of the regulatory mechanisms of T_{G1} to cell external cues which provides the link between the proliferative behavior of the individual cells and the general histogenetic program that is dependent upon the coordinate proliferative behavior of vast numbers of cells that constitute the overall PVE.

G1 phase is divisible into an initial phase, when its regulatory mechanisms are responsive to cell-external influences, and a second phase, some one to two hours prior to the G1/S transition, when its regulatory mechanisms are not responsive to cell-external influences (Coats et al. 1996; Massague and Polyak 1995). The transition between these two phases is referred to as the "restriction point" (Murray and Hunt 1993; Pardee 1989). Passage through the restriction point and subsequently through the G1/S transition reflects the balance of mechanisms that act to drive the cell forward through G1 and those that act to arrest its progression before the restriction point (Coats et al. 1996; Massague and Polyak 1995). Advance of G1 reflects principally the consequences of activation of cyclin dependent kinases (cdk's). Such activation depends upon the formation of cdk complexes with regulatory protein subunits referred to as cyclins (Sherr 1993). Other modifications of the cdk's, in particular phosphorylation by cdk-activating kinases (CAK's), are also required (Sherr 1994). Arrest of cycle progression in G1 reflects principally the action of a variety of classes of proteins that down regulate the formation or activation of the cdk complexes (Elledge and Harper 1994; Hengst and Reed 1996).

The passage through the G1/S transition is especially dependent upon the kinase action of cdk2 (Tsai et al. 1991, 1993). The activity of this kinase depends in turn upon binding to the regulatory subunit cyclin E. Formation of the cdk2/ cyclin E complex, which also includes PCNA, a protein regulatory to DNA polymerization, is critically timed to the G1 to S phase transition (Ohtsubo et al. 1995; Tsai et al. 1993). The role of the complex appears to be specific to the G1/S transition (Koff et al. 1992, 1993). In dissociated cell systems *in vitro*, cyclin E expression is largely delayed beyond the restriction point, and its peak expression is fine tuned to the G1/S transition (Lew et al. 1991).

Transition through the restriction point is inhibited by multiple classes of cell internal proteins that act to oppose the progression of G1 beyond this point (Coats et al. 1996; Elledge and Harper 1994; Sherr and Roberts 1995). Among these are members of at least two families of "cycle inhibitors," as well as the Rb and the p53 proteins. The cycle inhibitors, p27 (Hengst and Reed 1996; Sherr and Roberts 1995) and p21 (de Nooij and K 1995; Massague and Polyak 1995) are closely related with respect to structure and function. The action of p21 involves, at least in part, down regulation of formation of the cyclin E/cdk2 complex (Massague and Polyak 1995).

Down regulation of the effect of the inhibitory proteins to allow G1 progression occurs before the restriction point during G1 phase (Elledge and Harper 1994; Massague and Polyak 1995; Sherr 1994). Up regulation of members of a family of D cyclins is critical to this, both because the cyclin D proteins down regulate the action of the specific inhibitor proteins and because they promote the phosphorylation and thereby deactivation of the cycle inhibitory Rb. TGF-β-induced cycle blockade appears to act by down regulating the formation of the D cyclins with up regulation of the G1 inhibitors, stalling the phosphorylation of Rb. These are all prerestriction point effects (Massague and Polyak 1995; Parker et al. 1995). Secondarily, however, there is disruption of the formation of the cdk2/cyclin E complex with consequences for the G1/S transition after the restriction point (Massague and Polyak 1995).

T_{G1} Regulation and Exit from Cycle

FGF-β is only one example of the cell external substances that act to modulate the progression of G1. For the reasons reviewed above, it is likely that all act via the cell-internal, pre-restriction point mechanisms of regulation of G1 progression. In dissociated cells from the PVE, β-FGF has been determined to be "mitogenic." That is, it holds cells in cycle and drives proliferation (Ghosh and Greenberg 1995; Kilpatrick and Bartlett 1993). In other words it decreases Q. It is our hypothesis that mitogenic substances drive proliferative activity by holding cells in cycle and also act to decrease T_{G1}. We suggest that the mechanism involves down regulation of the G1 inhibitory system with accelerated formation of the cdk2/cyclin E complex. As precedent for this hypothesis, it has been established in dissociated cell systems that transgene enforced overexpression of both cyclin E and D not only drives the proliferative activity of cells but also decreases T_{G1} (Ohtsubo and Roberts 1993; Quelle et al. 1993; Resnitzky et al. 1994). Further, the degree of shortening of G1 is proportional to the level of cyclin E overexpression (Ohtsubo et al. 1995).

NT-3 (Ghosh and Greenberg 1995), like TGF-β (Massague and Polyak, 1995), on the other hand, is "anti-mitogenic." That is, it drives cells out of cycle and brakes proliferation. In other words it increases Q. It is our hypothesis that anti-mitogenic substances brake proliferative activity by increasing the probability that cells will exit the cycle and be directed along a path of postmitotic differentiation. This consequence, we predict, will be associated with increase in T_{G1}. We predict, further, that this effect of antimitogenic substances will involve up regulation of the G1 inhibitory system with decelerated formation of the cdk2/cyclin E complex. Consistent with this expectation, we have demonstrated that there is an inverse relationship between T_{G1} and the quantitative expression of message for cyclin E in the murine PVE (Delalle et al. 1995 and unpublished observations). Expression of message for p21 and cdk2, by contrast to that of cyclin E, appears not to be regulated through the cell cycle in that expression for both is uniformly distributed through the full width of the PVE (Delalle and Caviness 1996 and unpublished observations). The findings regarding p21 and cdk2 are consistent with patterns of message expression for these two proteins in dissociated cell systems *in vitro* where it appears that their activities are regulated post-transcriptionally rather than by synthesis (Hengst and Reed 1996; Koff et al. 1992). Glutamate (via kainate receptor) and GABA (via the GABA-A receptor) also "decrease DNA synthesis" in the PVE (Lo Turco et al. 1995), possibly by mechanisms similar to those of NT-3 or TGF-β. To the extent that our formulation is valid, Q, which is the the probability of cell cycle exit, is obligatorily linked to T_{G1}. Because Q varies coordinately with T_{G1}, we suggest that the probability for cycle exit bears a specific relation to the value of T_{G1}. That is, Q should increase as T_{G1} is prolonged, which is exactly what is observed (Fig. 4; Takahashi et al. 1996 b).

T_{G1} and the Transcription Profile

The coordinate relationship of T_{G1} and Q proposed here has central significance for regulation of the number of cells produced. The significance of this coordinate relationship may extend to important aspects of histogenesis other than the regulation of neuronal output in the strict numerical sense. We suggest, as an example, that the pattern of advance of T_{G1} over the neuronogenetic interval might play a role in specification of neuronal *class* (Fig. 9), whereas the linked value of Q would govern the *number* of neurons of each class which are produced (Fig. 8; Caviness et al. 1995). This phenomenon gains in theoretical interest from observations elsewhere that indicate that mechanisms regulatory to T_{G1} also shape the cell transcription profile (Hinds and Weinberg 1994; Sherr 1994; Weinberg 1991) and through this mechanism might act directly to influence the patterns of differentiation followed by cells after the leave the cycle.

With respect to regulation of T_{G1}, these influences on the transcription profile may be more or less direct (Segal and Greenberg 1996). The more direct would relate to the process of phosphorylation of Rb (and likely others of the cycle inhibitors) in preparation for the G1/S transition. This action releases a variety of transcription factors otherwise sequestered by the cycle inhibitory proteins (Hinds and Weinberg 1994; Sherr 1994; Weinberg 1991). Whereas some of the released transcription factors are probably dedicated to requirements of the G1/S transition, others probably have functions not related to cycle progression and possibly related to the differentiated functions of the cell. It is also evident that growth factors that act mitogenically or antimitogenically (suggested above to be by shortening or lengthening G1, respectively) directly affect the cell transcriptional profile more directly via the Ras and other calcium release-dependent pathways (Segal and Greenberg 1996). In summary, the transcriptional profile specifying each of the sequence of neuronal classes or specifying architectonic map positions may be coordinately regulated with T_{G1}. The mechanisms of such regulation may be both direct through those that regulate T_{G1} itself and indirect as a parallel effect of a cell external, cycle modulator that also coordinately modulates T_{G1}.

Concluding Overview

The implications of the coordinate relationships between regulation of T_{G1} on the one hand, and cell output, determination of cell class and specification of neocortical map position on the other hand potentially provide linkage between the cell biological processes of histogenesis and the molecular biological processes of cell cycle regulation and gene transcription. These coordinate relationships imply that cell external and internal mechanisms regulating T_{G1} for any given integer cell cycle are directly or indirectly the same mechanisms that set the transcription profile of cells emerging from the PVE with that cycle. They imply that step advances in T_{G1} are coordinate with systematic and specific step alteration in the panel of genes transcribed. This panel may be expected to change stepwise as the neuronogenetic process advances through the succession of neuronal classes that constitute the neocortical layers. In principle these hypotheses could be tested experimentally by manipulations that disrupt in a controlled and time-limited way the cell internal mechanisms of regulation of T_{G1}.

Acknowledgments

Supported by NIH grants NS12005 and NS28061 and NASA grant NAG2-950. T.T. was supported by a Fellowship of The Medical Foundation, Inc., Charles A. King Trust, Boston, MA. Valuable discussions with Pradeep Bhide, Sahoko Miyama are gratefully acknowledged.

References

Austin CP, Cepko CL (1990) Cellular migration patterns in the developing mouse cerebral cortex. Development 110:713–732

Bayer SA, Altman J (1991) Neocortical development. New York, Raven Press, Inc.

Bennett MVL, Barrio LC, Bargiello TA, Spray DC, Hertzberg EC, Saez JC (1991) Gap junctions: new tools, new answers, new questions. Neuron 6:305–320

Bisconte J.-C, Marty R (1975) Analyse chronoarchitectonique du cerveau de rat par radioautographie. I. Histogenese du telencephale. J Hirnforsch 16:55–74

Blaschke A, Staley K, Chun J (1996) Widespread programmed cell death in proliferative and postmitotic regions of the fetal cerebral cortex. Development 122:1165–1174

Caviness VS, Jr (1982) Neocortical histogenesis in normal and reeler mice: a developmental study upon [3H]thymidine autoradiography. Dev Brain Res 4:293–302

Caviness VS, Jr (1988) Architecture and development of the thalamocortical projection in the mouse. In: Bentivoglio M, Spreafico R (eds) Cellular thalamic mechanisms, Amsterdam, Elsevier Science Publishers, pp. 489–499

Caviness VS, Jr, Takahashi T, Nowakowski RS (1995) Numbers, time and neocortical neuronogenesis: A general developmental and evolutionary model. Trends Neurosci 18:379–383

Coats S, Flanagan WM, Nourse J, Roberts JM (1996) Requirement of $p27^{Kip1}$ for restriction point control of fibroblast cell cycle. Science 272:877–880

de Nooij JC, K, HI (1995) Uncoupling cell fate determination from patterned cell division in the *Drosophila* eye. Science 270:983–985

Delalle I, Caviness VS, Jr (1996) Temporal and spatial expression of $p21^{Cip1/WAF1}$ in the developing murine neocortical cerebral wall. Soc Neusci Abst 22:524

Delalle I, Tsai L.-H, Harlow E, Caviness VS, Jr (1995) Temporal and spatial expression of cyclins B and E in the developing murine cortex. Soc Neurosci Abst 21:287

Dermietzel R, Spray DC, ((1993) Gap junctions in the brain: where, what type, how many and why? Trends Neurosci 16: 186–192

Elledge S and Harper W (1994) Cdk inhibitors: on the threshold of checkpoints and development. Curr Opin Cell Biol 6:847–852

Erzurumlu RS, Jhaveri S (1992) Emergence of connectivity in the embryonic rat parietal cortex. Cereb Cortex 2:336–352

Fernandez V, Bravo H (1974) Autoradiographic study of the cerebral cortex in the rabbit. Brain Behav Evol 9:317–332

Ferrar I, Soriano E, Del Rio JA, Alcantara S, Auladell C (1992) Cell death and removal in the cerebral cortex during development. Prog Neurobiol 39:1–43

Finlay BL, Pallas SL (1989) Control of cell number in the developing mammalian visual system. Prog Neurobiol 32:207–234

Fujita S (1960) Mitotic pattern and histogenesis of the central nervous system. Nature 185:702–703

Ghosh A, Greenberg ME (1995) Distinct roles for bFGF and NT-3 in the regulation of cortical neurogenesis. Neuron 15:89–103

Goodwin BC, Cohen MH (1969) A phase-shift model for the spatial and temporal organization of developing systems. J Theoret Biol 25:49–107

Goto T, Takahashi T, Bhide PG, Miyama S, Caviness VS, Jr (1996) b-FGF as cell cycle modulator in vitro in the neocortical proliferative epithelium of embryonic mice. Soc Neurosci Abst 22:523

Hengst L, Reed SL (1996) Translational control of $p27^{Kip1}$ accumulation during the cell cycle. Science 271:1861–1864

Heumann D, Leuba G, Rabinowicz T (1978) Postnatal development of the mouse cerebral neocortex IV. Evolution of the total cortical volume of the population of neurons and glial cells. J Hirnforsch 19:411–416

Hicks SP, D'Amato CJ (1968) Cell migration to the isocortex in the rat. Anat Rec 160:619–634

Hinds JW, Ruffett TL (1971) Cell proliferation in the neural tube: an electron microscopic and Golgi analysis in the mouse cerebral vesicle. Z Zellforsch 115:226–264

Hinds PW, Weinberg RA (1994) Tumor suppressor genes. Curr Opin Genet Dev 4:135–141

Kilpatrick TJ, Bartlett PF (1993) Cloning and growth of multipotential neural precursors: requirements for proliferation and differentiation. Neuron 10:255–265

Koff A, Giordano A, Desai D, Yamashita K, Harper JW, Elledge S, Nishimoto T, Morgan DO, Franza BR, Roberts JM (1992) Formation and activation of a cyclin E-cdk2 complex during the G1 phase of the human cell cycle. Science 257:1689–1694

Koff A, Ohtsuki M, Polyak K, Roberts JM, Massague J (1993) Negative regulation of G1 progression in mammalian cells; inhibition of cyclin E-dependent kinase by TGF-β. Science 260:536–539

Lavdas A, Mione MC, Parnavelas JG (1996) Neuronal clones in the cerebral cortex show morphological and neurotransmitter heterogeneity during development. Cereb Cort 6:490–497

Lew DJ, Dulic V, Reed SI (1991) Isolation of three novel human cyclins by rescue of G1 cyclin (Cln) function in yeast. Cell 66:1197–1206

LoTurco JJ, Kriegstein A (1991) Clusters of coupled neuroblasts in embryonic neocortex. Science 252:563–566

LoTurco JJ, Owens DF, Heath MJS, Davis MBE, Kriegstein AR (1995) GABA and glutamate depolarize cortical progenitor cells and inhibit DNA synthesis. Neuron 15:1287–1298

Massague J, Polyak K (1995) Mammalian antiproliferative signals and their targets. Curr Opin Gen Dev 5:91–96

McConnell SK (1991) The generation of neuronal diversity in the central nervous system. In: Cowan WM, Shooter EM, Stevens CF, Thompson RF (eds), Ann Rev Neurosci, Palo Alto, Annual Reviews, Inc, pp. 269–300

Meinhardt H (1982) Models of biological pattern formation. New York, Academic Press

Miller MW (1985) Cogeneration of retrogradely labeled corticocortical projection and GABA-immunoreactive local circuit neurons in cerebral cortex. Dev Brain Res 23:187–192

Misson J.-P, Austin CP, Takahashi T, Cepko CL, Caviness VS, Jr (1991) The alignment of migrating neural cells in relation to the murine neopallial radial glial fiber system. Cereb Cort 1:221–229

Miyama S, Bhide P, Takahashi T, Caviness VS, Jr (1995) Striatum modulates neocortical cytogenesis in mice. Soc Neurosci Abst 21:286

Miyama S, Takahashi T, Nowakowski RS, Caviness VS, Jr (1997) A gradient in the duration of the q-phase in the murine neocortical proliferative epithelium. Cereb Cortex in press

Molnar Z, Blakemore C (1992) How are thalamocortical axons guided in the reeler mouse? Soc Neurosci, Abstr 18:778

Murray A, Hunt T (1993) The cell cycle. New York, WH Freeman and Co

Murray A, Kirschner MW (1991) What controls the cell cycle? Sci Am 264:56–63

Ohtsubo M, Roberts JM (1993) Cyclin-dependent regulation of G1 in mammalian fibroblasts. Science 259:1908–1912

Ohtsubo M, Theodoras AM, Schumacher J, Roberts JM, Pagano M (1995) Human cyclin E, a nuclear protein essential to the G1-to-S phase transition. Mol Cell Biol 15:2612–2624

O'Leary DDM, Schlaggar BL, Tuttle R (1994) Specification of neocortical areas and thalamocortical connections. In: Cowan WM, Shooter EM, Stevens CF, Thompson RF (eds) Annual Rev Neurosci Palo Alto, CA: Annual Reviews, Inc, pp. 419–439

Pardee AB (1989) G1 events and regulation of cell proliferation. Science 246:603–608

Parker SB, Eichele G, Zhang P, Rawls A, Sands AT, Bradley A, Olson EN, Harper JW, Elledge SJ (1995) p53-Independent expression of $p21^{Cip1}$ in muscle and other terminally differentiating cells. Science 267:1024–1027

Parnavelas J, Barfield JA, Franke E, Luskin MB (1991) Separate progenitor cells give rise to pyramidal and nonpyramidal neurons in the rat telencephalon. Cereb Cort 1:463–468

Peters A, Jones EG (1984) Classification of cortical neurons. New York, Plenum Press

Quelle DE, Ashmun RA, Shurtleff SA, Kato J.-Y, Bar-Sagi D, Roussel MF, Sher CJ (1993) Overexpression of mouse D-type cyclins accelerate G1 phase in rodent fibroblasts. Genes Dev 7:1559–1571

Rakic P (1974) Neurons in rhesus monkey visual cortex: systematic relation between time of origin and eventual disposition. Science 183:425–427

Rakic P (1988) Specification of cerebral cortical areas. Science 241;170–176

Resnitzky D, Gossen M, Bujard H, Reed SI (1994) Acceleration of the G1/S phase transition by expression of cyclins D1 and E with an inducible system. Mol Cell Biol 14:1669–1697

Sauer FC (1936) The interkinetic migration of embryonic epithelial nuclei. J Morphol 60:1–11

Segal RA, Greenberg ME (1996) Intracellular signaling pathways activated by neurotrophic factors. Ann Rev Neurosci 19:463–489

Sherr CJ (1993) Mammalian G1 cyclins. Cell 73:1059–1065

Sherr CJ (1994) G1 phase progression: cycling on cue. Cell 79:551–555

Sherr CJ, Roberts JM (1995) Inhibitors of mammalian G1 cyclin-dependent kinases. Genes Devel 9:1149–1163

Sidman RL, Rakic P (1973) Neuronal migration, with special reference to developing human brain: a review. Brain Res 62:1–35

Smart IHM, McSherry GM (1982) Growth patterns in the lateral wall of the mouse telencephalon. II. Histological changes during and subsequent to the period of isocortical neuron production. J Anat 131:415–442

Smart IHM, Smart M (1982) Growth patterns in the lateral wall of the mouse telencephalon. I. Autoradiographic studies of the histogenesis of the iso-cortex and adjacent areas. J Anat 134:273–298

Stensaas LJ, Stensaas SS (1968) An electron microscope study of cells in the matrix and intermediate laminae of the cerebral hemisphere of the 45 mm rabbit embryo. Z Zellforsch 91:341–365

Takahashi T, Goto T, Miyama S, Nowakowski RS, Caviness VS, Jr (1996a) Intracortical distribution of a cohort of cells arising in the PVE. Soc Neurosci Abst 22:284

Takahashi T, Nowakowski RS, Caviness VS, Jr (1992) BUdR as an S-phase marker for quantitative studies of cytokinetic behaviour in the murine cerebral ventricular zone. J Neurocytol 21:185–197

Takahashi T, Nowakowski RS, Caviness VS, Jr (1993) Cell cycle parameters and patterns of nuclear movement in the neocortical proliferative zone of the fetal mouse. J Neurosci 13:820–833

Takahashi T, Nowakowski RS, Caviness VS, Jr (1994) Mode of cell proliferation in the developing mouse neocortex. Proc Natl Acad Sci USA 91:375–379

Takahashi T, Nowakowski RS, Caviness VS, Jr (1995) The cell cycle of the pseudostratified ventricular-epithelium of the murine cerebral wall. J Neurosci 15:6046–6057

Takahashi T, Nowakowski RS, Caviness VS, Jr (1996b) Interkinetic and migratory behavior of a cohort of neocortical neurons arising in the early embryonic murine cerebral wall. J Neurosci 16:5762–5776

Takahashi T, Nowakowski RS, and Caviness VS, Jr (1996c) The Q fraction of the murine cerebral PVE and a general computational model of neocortical neuronogenesis. J Neurosci 16:6183–6196

Tsai L.-H, Harlow E, Meyerson M (1991) Isolation of the human cdk2 gene that encodes the cyclin A- and adenovirus E1A- associated p33 kinase. Nature 353:174–177

Tsai L.-H, Lees E, Faha B, Harlow E, Riabowol K (1993) The cdk2 kinase is required for the G1-to-S transition in mammalian cells. Oncogene 8:1593–1602

Weinberg RA (1991) Tumor suppressor genes. Science 254:1138–1146

Wolpert L (1969) Positional information and the spatial pattern of cellular differentiation. J Theoret Biol 25:1–47

Wolpert L (1978) Gap junctions: channels for communication in development. In: Feldman J, Gilula NB, Pitts JD (eds) Intercellular junctions and synapses (Receptors and recognition, Series B). London, Chapman and Hall, pp. 83–96

Wood JG, Martin S, Price DJ (1992) Evidence that the earliest generated cells of the murine cerebral cortex form a transient population in the subplate and marginal zone. Dev Brain Res 66:137–140

The Nature and Nurture of Cortical Development

H. Kennedy and C. Dehay

Introduction

Damage to the cerebral cortex in man leads to a wide range of cognitive, sensory and motor deficits and for the proverbial "man in the street" many of the attributes loosely associated with the brain are in fact characteristic of cortical function. Given the complexity of its function and its pre-eminence in man, the cerebral cortex has an apparently simple basic composition. This simplicity is not only reflected by a relatively small number of cell types but also in its gross architectural organization. The cortex is subdivided tangentially into four to eight layers. Regional variations of the number of layers and the density of neurones in each layer allowed early pioneers of cortical anatomy to distinguish cortical areas (Brodmann 1909), which were rapidly thought of as housing specific cognitive, sensory or motor functions (Payne et al. 1996). Within the visual system, each area was envisioned to process a particular aspect of visual information (Zeki 1978).

It is now commonly accepted that there is a varying degree of localization of function, so that although we spoke glibly in the 1980s of the color area, it remains true that as our understanding of cortical physiology improves, we can relate individual cortical areas with a restricted range of specific functions (Payne et al. 1996).

The uniqueness of the physiological function that characterises each cortical area is thought to be derived from the specificity of its extrinsic connectivity rather than from important differences in the microcircuitry of each cortical area (Douglas et al. 1988). The cortical layers exhibit a stereotyped connectivity and display a number of common features across areas, so that, for example, the major thalamic input to the cortex is to the middle layers and the major output to midbrain and thalamus is from lower layers, whereas the principal connectivity between cortical areas is from both upper and lower layers.

These considerations allow one to think of the organization of the cortex as being composed of two separate and interdependent domains: cortical layers and cortical areas. This view of the cortex brings hope to the student of cortical development because there is evidence that relatively well-defined and distinct mechanisms control the development of each of these two domains (Shatz 1992).

Control of cortical development has been thought of as being of two origins. Projections from outside the cortex have been shown to exert a determining

A. M. Galaburda / Y. Christen (Eds.)
Normal and Abnormal Development of Cortex
© Springer-Verlag Berlin Heidelberg New York 1997

influence, so that it has been suggested that the development of regional features is imposed by the periphery (O'Leary 1989; Killackey 1990). This concept has exerted a certain fascination since it suggests that the complexity of the cortical function is made possible by a prolonged developmental period in which environmental signals shape the emerging circuitry of the cortex. Here environmental signals are often loosely defined to include not only local cellular signalling but also ascending activity, ultimately derivable from the environment of the organism (Purves and Lichtman 1985). This epigenetic control of cortical development was initially envisioned to palliate a degree of complexity that was thought to be too great to be specified by the finite number of genes contained in the genome (Changeux and Danchin 1976; Changeux 1983). Alternatively, cortical development is thought of as being largely determined by a control that is intrinsic to the developing cortex. According to this theory, regional specificity will be generated independently and perhaps prior to the arrival of peripheral projections and will come under genetic control.

Historically, intrinsic and extrinsic control of development were viewed as controlling independently cortical layers and areas, respectively (Shatz 1992). The involvement of intrinsic control of layer formation was largely derived from experiments showing that commitment of neuroblasts to a particular layer occurs prior to neuronal migration so that laminar fate is determined in the ventricular zone (Caviness and Sidman 1973; Caviness 1982; McConnell 1991; McConnell and Kaznowski 1991; Rakic 1974, 1976). The importance of extrinsic control of cortical area identity was derived from intra-cortical graft experiments (Schlaggar and O'Leary 1991) and the experimental manipulation of areal borders (Rakic 1988; Rakic et al. 1991; Rhoades et al. 1993; Dehay et al. 1996 a).

This dichotomy, however, has turned out to be too good or too simple to be true. Recent results including intra-graft experiments confirm that the regional connectivity of cortical areas show some autonomy (Arimatsu 1994; Arimatsu et al. 1992, 1994; Barbe and Levitt 1992; Cohen-Tannoudji et al. 1994; Ebrahimi-Gaillard et al. 1994; Ebrahimi-Gaillard and Roger 1996). These and other findings bring home a basic truth: cortical areas are composed of layers and one can expect a coupling between the developmental control of these two cortical domains.

It has been suggested that intrinsic and extrinsic control are not alternative mechanisms (Rakic 1988; Kennedy and Dehay 1993 a). Adult features that are achieved early in development do so independently of late environmental factors and at most can be perturbed by them, so that one can conclude that these factors do not play an instructive role. On the other hand, patterns that are achieved late in development may depend on environmental regulation. In this way it is crucially important to determine the exact developmental stage when a particular organizational feature emerges during the prolonged period that is the hallmark of cortical development in the primate (Kennedy and Dehay 1993 b). In the present review we shall focus on recent findings in our laboratory which suggest that early-formed areal patterns, both of neurone, number and the precision of connections, are a conspicuous feature of primate development. These findings contrast with a major late-developmental event that is highly specific to feedback cortical pathways and further appears to be largely characteristic of the primate. This remodelling of feedback connections in sensory areas of the primate could play an important role in the neurocognitive development of this order.

Emerging Patterns in Cytoarchitectonics

Theories of cortical function have alluded to the notion of a homogeneous organization in which a basic canonical circuit is repeated lattice-like (Mountcastle 1957; Hubel and Wiesel 1977; Douglas et al. 1988). Stereological measurements of neurone number partly support this notion of homogeneity, since it has been reported that the cortical thickness of the cortex is roughly constant both across areas and across species (Rockel et al. 1980). Accordingly, the differences in the cytoarchitectonics that characterise cortical areas must reflect *local* differences in numbers and densities of neurones. One possibility is that the early-formed cortex contains a more even allocation of neurones in individual layers and that high levels of cell death, which are characteristic of the developmental period, occur to a varying extent in different layers. Work in this area showed that differential cell death indeed does play a role in the emergence of areal cytoarchitectonics of the adult (Finlay and Slattery 1983; Finlay and Pallas 1989). However, could it be that these different levels of cell death are acting upon areal differences in the numbers of neurones established at the moment of neurogenesis, as has been shown in the embryonic spinal cord (Oppenheim et al. 1989)? Birthdating experiments are compatible with this possibility. Pulse injections of tritiated thymidine (3H-thy) show that there are significant areal differences in the laminar distribution of labeled neurones in the monkey neocortex (Rakic 1982; Dehay et al. 1993) as well as in the hippocampus of mouse and monkey (Angevine 1965; Rakic and Nowakowski 1981). However, although these findings could be interpreted as resulting from regional differences in neurone output of the ventricular zone, they could equally be generated by local differences in levels of cell death. Further, modern molecular techniques show levels of cell death in excess of 50% in developing cortex (Blaschke et al. 1996), so that small regional variations of this regressive phenomenom could easily generate the observed variations of 3H-thy labeled neurones.

To investigate this issue we have carried out birthdating experiments in fetal mice (Polleux et al. 1995; 1997). Pulse injections of 3H-thy were used to describe the inside-first outside last sequence of layer production. When the laminar distribution of labeled neurones was compared in fronto-parietal areas, important differences in the laminar distribution of labeled neurones were found, similar to what had previously been described in monkey visual cortex (Dehay et al. 1993; Kennedy et al. 1996b). To exclude the influence of cell death on the results, we have examined the differences in the percentages of labeled cells with respect to the total number of neurones (defined as the labelling index) in homologous layers of adjacent areas (Fig. 1). Because cell death will affect labeled and unlabeled

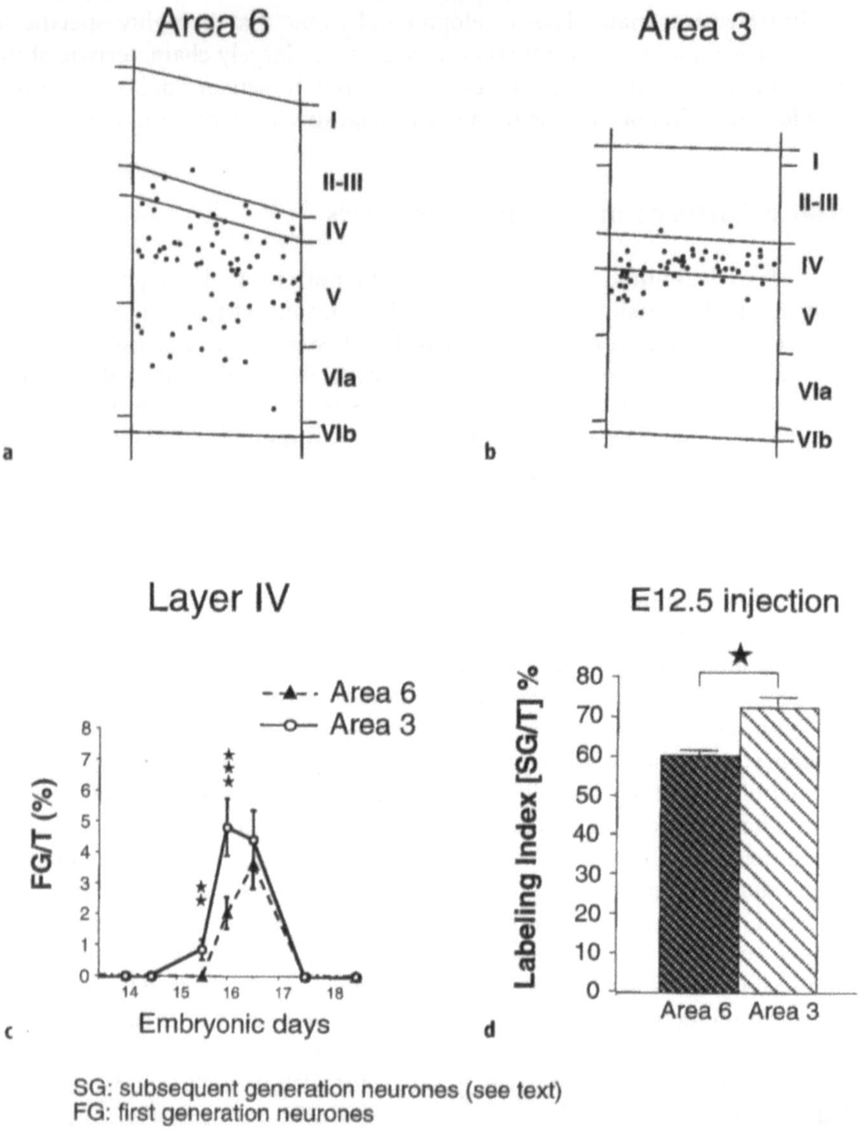

Fig. 1. Areal differences in the timetable of layer production. Injections of 3H-thymidine (E16.0) show significant differences in the laminar distribution of labeled neurones in neighbouring cortical areas in the adult (**a** and **b**). By computing the number of heavily labeled first generation (FG) neurones in a cortical layer we can generate laminar generation curves that describe the timetable of layer production in each area (**c**). These differences in the timetable of layer production could be generated by differences in the recruitment of precursors in the generation of a layer (Frantz and McConnell 1996). By comparing SG/T in areas 3 and 6 immediately above FG neurones we can conclude that the labeling indices in the precursors generating these two areas differ (**d**). This shows that precursors committed to the generation of a layer have different cell-cycle kinetics, which likewise contribute to areal differences in rates of production of homologous layers

neurones equally, true differences in labeling index of homologous layers will reflect differences in the rate of generation of these layers. This quantitative approach makes it possible to reliably compare the timetable of layer production across areas and shows that there are important areal differences in the onset, duration and cessation of infragranular layer production in adjacent fronto-parietal areas in the rodent (Polleux et al. 1995; 1997).

These results showing areal differences in the timetable of layer production indicate an important coupling between the developmental control of layers and areas. They indicate that, at a given stage of corticogenesis, precursors separated by only several hundred microns will be generating neurones that are destined for different layers and will form different sorts of connections. The work of Susan McConnell (McConnell 1988; McConnell and Kaznowski 1991; Frantz and McConnell 1996) has shown that cortical precursors are multipotential and that the commitment to a given cortical layer is made shortly after the final S-phase in the last round of mitosis, presumably in response to some sort of environmental signal. Is there any possibility that the kinetics of this final round of mitosis could contribute to the numbers of neuroblasts destined for a given layer? If this were so it would mean that there is a regionalisation of the kinetics of the cell-cycle. Numerous workers have used labeling indices as an indication of cell-cycle. It is commonly accepted that S-phase (the phase of DNA synthesis) has a constant duration throughout corticogenesis, whereas the duration of the cell-cycle (Tc) becomes progressively longer at later stages of corticogenesis (Waechter and Jaensch 1972; Schultze and Korr 1981; Schmahl 1983; Miller and Kuhn 1995; Takahashi et al. 1995). Hence, in a pool of precursors, a 3H-thy pulse will label progressively a smaller fraction of precursor as corticogenesis progresses. In this way differences in the labeling indices reflect differences in the cell-cycle duration (Fig. 2).

At fetal stages we have no morphological indication of areal limits in the ventricular zone of the rodent. This makes it necessary to examine adult material. If labeling resulting from fetal injections is examined at later stages, those precursors that were in S-phase at the moment of the pulse will give rise to labeled cortical neurones showing a wide range of labeling intensities. Those precursors that underwent their final division just after the 3H-thy pulse will lead to a class of heavily labeled neurones that correspond to the first generation (FG) of labeled neurones. Those labeled precursors that continued to divide will give rise to second, third, fourth, etc. generations of labeled neurones where the degree of labeling is halved at each round of mitosis and which are identified by their intensity as subsequent generation (SG) neurones. Theoretically, the labeling index in a restricted layer or sub-layer of the adult will reflect the labeling index that would have been observed in the ventricular zone if the material had been examined shortly after delivery of the pulse (Schultze et al. 1974; Schultze and Korr 1981). Once again, the procedure is not influenced by cell death, since labeled and unlabeled neurones will be equally affected by this phenomenon. However, FG neurones in mouse are distributed over two to three cortical layers so that the unlabeled neurones that are in the region where FG neurones are

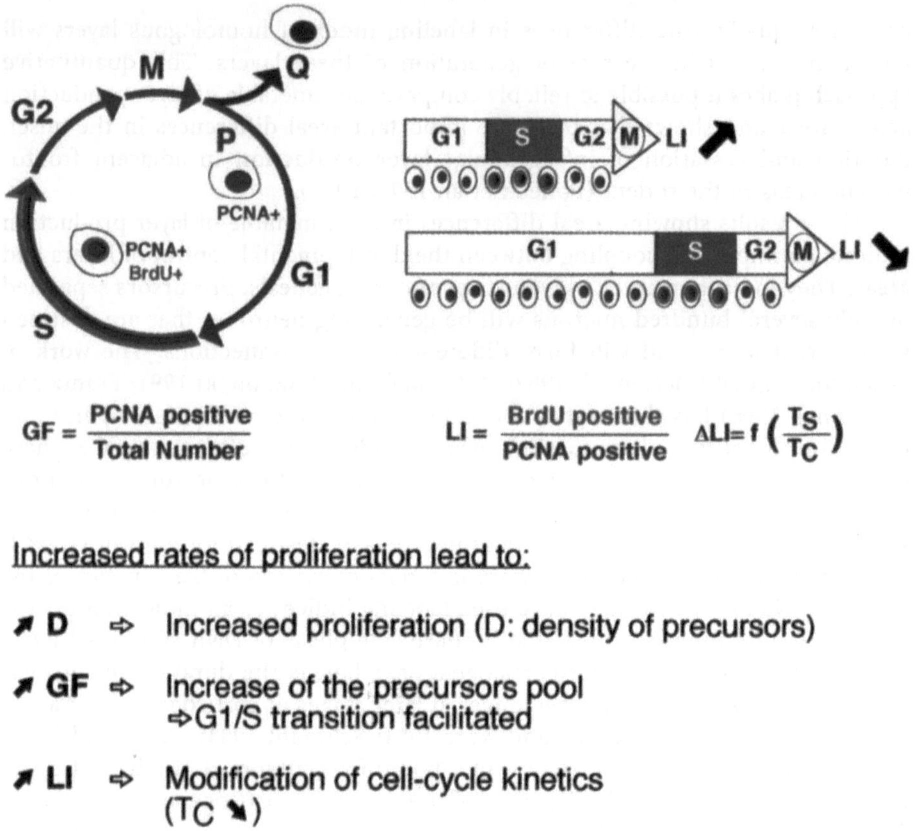

Fig. 2. Cell-cycle kinetics of cortical precursors. The duration of S-phase is thought to remain constant and changes in the overall duration of the cell-cycle (Tc) are usually the consequence of increasing G1 phase. One consequence of this is that a pulse of an S-phase marker (BrdU) will lead to a high percentage of labeled precursors (labeling index; LI) in a fast cycling population and a low LI in a slow cycling population. PCNA immunostaining can be used to identify those precursors that are cycling and enables us to define the growth fraction (GF). Ts: duration of S-phase

found could be generated before or after the 3H-thy pulse. One way of overcoming this problem is to examine the labeling index of SG neurones immediately above the region of FG-labeled neurones. By definition *all* of the neurones superficial to FG neurones are generated after the 3H-thy pulse, and the progeny of those precursors that were in S-phase will be labeled SG neurones and those in G1 and G2-M will be unlabeled neurones (Fig. 1). When this analysis is carried out in mouse cortex, the differences in timetable of infragranular layer productions are shown to be accompanied by differences in kinetics of the different pools of precursors generating the adjacent cortical areas (Polleux et al., manuscript in preparation).

These results in the mouse show that adjacent regions of the ventricular zone follow proliferative programmes with different timing and distinct kinetics. Can these variations in kinetics generate the laminar differences in neurone number (Beaulieu 1993; Skoglund et al. 1996) in rodent fronto-parietal cortex? To answer this question we need to model corticogenesis. Numerous authors using a variety of techniques have determined the cell-cycle parameters during corticogenesis. At the onset of neurone production the duration of the cell-cycle (Tc) is 8 hours and progresses to 20 hours by the end of corticogenesis (Waechter and Jaensch 1972; Schultze and Korr 1981; Schnahl 1983; Miller and Kuhn 1995, Takahashi et al. 1995). At early stages, when Tc values are minimal, all divisions lead to daughter cells that remain in the cell-cycle and lead to a steady increase in the pool of precursors. At the onset of cortical neurone production, a small proportion of precursors gives rise to daughters that loose their proliferative capacity and migrate out to the cortical plate. As corticogenesis proceeds, this fraction of differentiative divisions goes from 0 % prior to the onset of neurone production of layer 6 to supposedly a very high percentage (100 %?) at the end of neuron production of layer 2 (Rakic 1977, 1995; Takahashi et al. 1996). Clearly the rate of production of neurones at any given instant is governed by three factors: the numbers of precursors, Tc and the proportion of differentiative divisions. By constructing a dynamic model of layer production during corticogenesis (Fig. 3), we have been able to show that the experimentally measured regional variations in Tc values generate the known differences in the neurone number of individual layers (Moraillon et al. 1996).

These findings in the rodent support conclusions from work in primates which showed important areal differences in neurone production. We have shown, by using 3H-thy pulse injections combined with short survival times in the embryonic monkey, that differential rates of proliferation underlie the histogenesis of distinct cortical areas. The precursors of area 17 show a selective increase in the rate of division during the generation of supragranular layers, whereas area 18 precursors of the same layers are characterised by lower rates of proliferation (Dehay et al. 1993). These results reveal an early specification of a cortical area's identity that is underlied by a regional specialisation of the proliferative characteristics of the progenitor cells in the germinal zones, in agreement with the proposition of Rakic (Rakic 1982, 1988), who hypothesised that the ventricular zone contained a protomap of the future cortical areas. 3H-thy pulse injections followed by several days survival give indications on the rates of migration of neuroblasts from the ventricular zone to the cortical plate (Kennedy et al. 1996 b). These experiments show that there are higher rates of migration in area 17 compared to area 18 and suggest that common control mechanisms could be involved in migration and proliferation.

To determine the cellular mechanisms that underlie higher rates of cell division in area 17 precursors, we studied the cell-cycle kinetics in dissociated cell cultures (Kennedy and Dehay 1995; Savatier et al. 1995). Cultured monkey area 17 precursors do continue to show higher rates of proliferation than do cultured area 18 precursors, suggesting that the cell-cycle dynamics characterising both sets of precursors are, to a certain extent, cell autonomous. Furthermore, by using a battery of *in vitro* techniques, including 1) flow cytometry analysis of

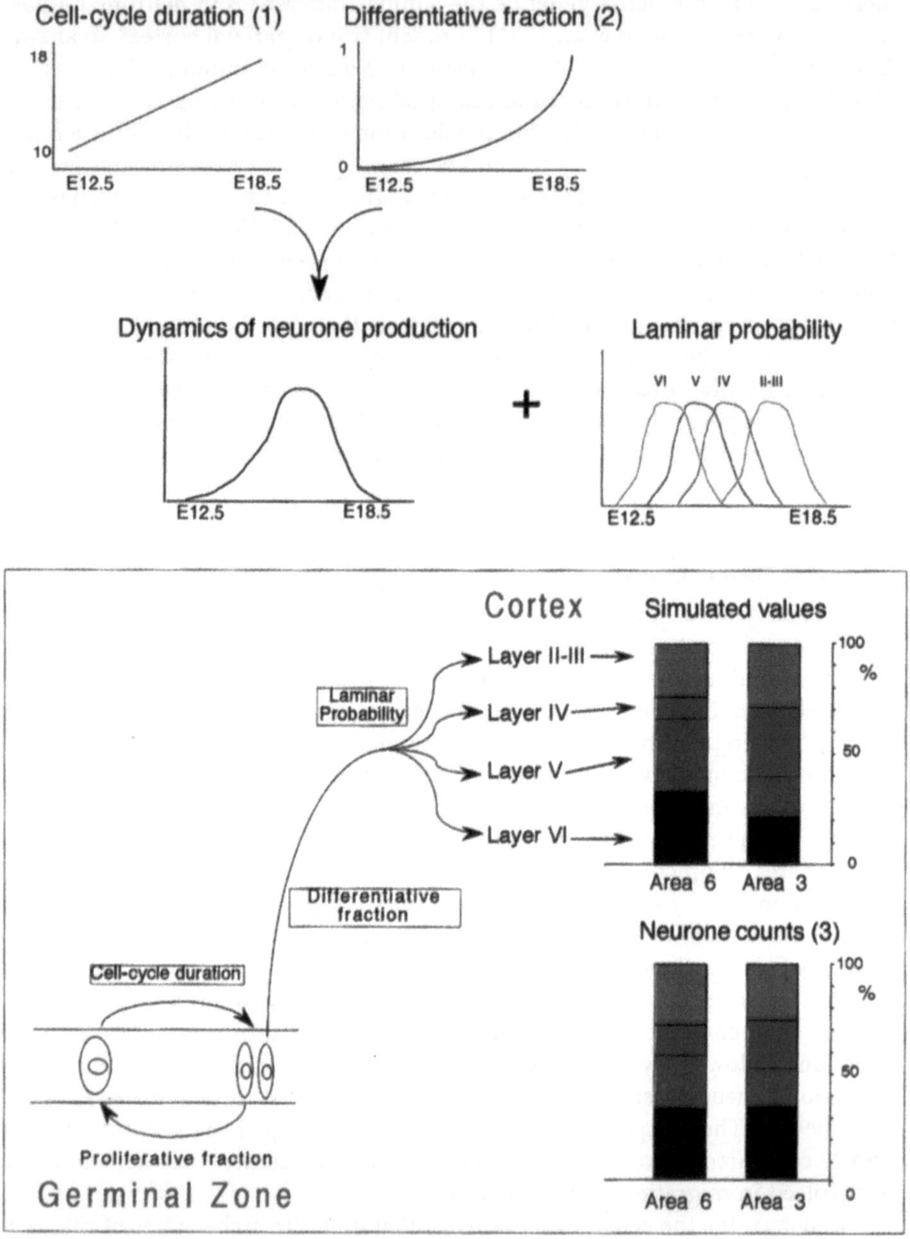
Cell-cycle duration (1)
18
10
E12.5
E18.5
Differentiative fraction (2)
1
0
E12.5
E18.5
Dynamics of neurone production
E12.5
E18.5
+
Laminar probability
VI
V
IV
II-III
E12.5
E18.5
Cortex
Simulated values
Layer II-III
Layer IV
Layer V
Layer VI
Laminar Probability
100
%
50
0
Area 6
Area 3
Differentiative fraction
Neurone counts (3)
100
%
50
0
Area 6
Area 3
Cell-cycle duration
Proliferative fraction
Germinal Zone

◀ **Fig. 3. Modelling of laminar neurogenesis in the specification of cortical areas.** The dynamics of neurone production is governed by the duration of the cell-cycle and the proportion of differentiative divisions. The developmental changes of these two parameters from the beginning of corticogenesis at E12.5 to the end of corticogenesis at E18.5 are now well documented and, together with the probability of layer production, determine the rate of laminar neurogenesis. The probability of layer production is based on pulse 3H-thymidine birthdating experiments in which the changing fraction of labeled neurones in individual cortical layers allows us to determine the duration and rate of laminar neurogenesis (Polleux et al. 1995; 1997). Computer simulation of these parameters using a dynamic compartment model allows us to compare theoretical numbers of neurones per layer with values obtained by stereological measurements. This procedure shows that the important differences in the numbers of neurones per cortical layer which characterise areas 3 and 6 can be largely explained by areal differences in the timetable of layer production. The small differences between the experimental and theoretical values suggest that cell death contributes to the final determination of neurone number in the cortex (Finlay and Slattery 1983). (1) Schmahl (1983); (2) Takahashi et al. (1996); (3) Skoglund et al. (1996)

propidium iodide and BrdU double labelled precursors, 2) labelled mitotic figures determination following colcemide-induced mitotic blockade and 3) BrdU cumulative labelling experiments, we have been able to measure the duration of individual phases of the cell-cycle for both sets of precursors (Fig. 2). These experiments show that Tc in monkey is extremely long – nearly 10 times the values in rodent. Hence, in the generation of supragranular layers, area 17 precursors show Tc values of about 169 hours and area 18 precursors, 191 hours (Savatier et al. 1995). The results showed that the shorter Tc of area 17 precursors is due to a reduced G1 phase in these cells, whereas S, G2 and M phases are similar in both sets of precursors. These findings show that the cellular mechanisms that underlie the early specification of cortical areas identity involve a fine regulation of the cell-cycle kinetics which is exerted during the G1 phase.

However, it should be kept in mind that, although these results provided the first and unambiguous demonstration that the identity of cortical areas is determined at the level of the germinal zone by specific properties of the precursors cells, they did not exclude the possible influence of environmental factors. One main environmental component of the embryonic cortex is the thalamocortical projection, which in the primate is in the vicinity of the germinal layers during a protracted period (Kennedy and Dehay 1993 a, b, c; Honig et al. 1996).

Prenatal enucleation experiments in the monkey (Dehay et al. 1989, 1991, 1996 a; Kennedy and Dehay 1993 a; Rakic 1988) show that early removal of geniculate afferents leads to a massive reduction in the numbers of LGN neurons, as well as to a drastic reduction of surface area of area 17 (Fig. 4). This effect is age dependent, so that enucleation performed during the early stages of cortical neurogenesis gives rise to a 75 % reduction of area 17 whereas deafferentation performed at later stages of neurogenesis results in a much more modest reduction (Dehay et al. 1991). In these early enucleates there is a reduction in the dimensions of area 17 but there is no overall change in the areal dimension of the neocortex (Dehay et al. 1996 a). This means that cortex that was originally destined to be area 17 has taken on the appearance of extrastriate cortex with its typical

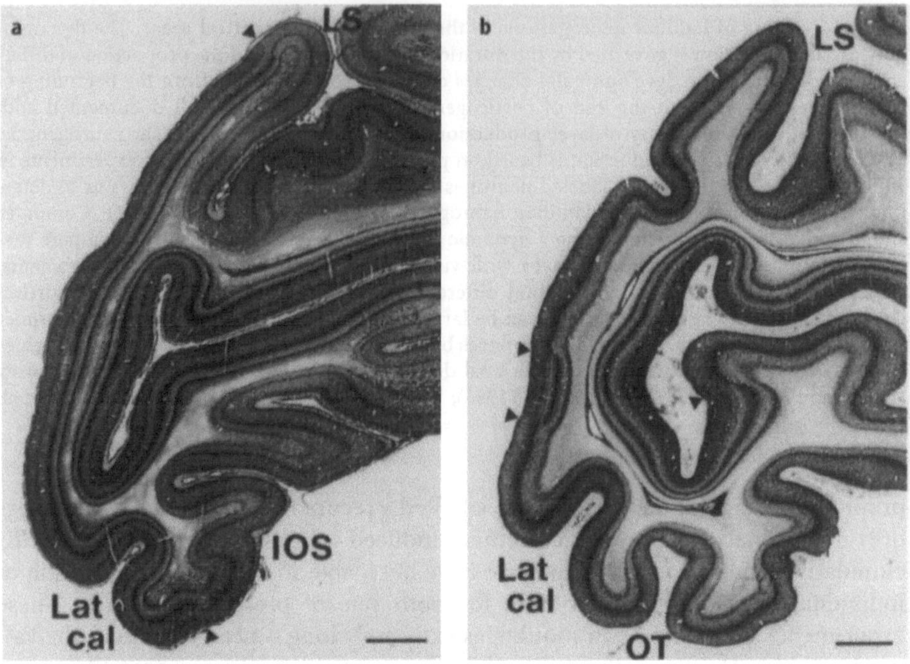

Fig. 4. Early bilateral enucleation in the fetus reduces the dimensions of striate cortex. Parasagittal sections of the occipital lobe from equivalent medio-lateral levels in normal neonate (**a**) and in a neonate that had undergone intrauterine surgery to remove both eyes on embryonic day 68 (**b**). Early bilateral enucleation induces a dramatic reduction of the lateral geniculate nucleus (not shown) and a reduction of the areal dimensions of striate cortex which is particularly pronounced on the operculum. Note that the lamination of the reduced striate cortex appears unaffected and the frontier with area 18 is as sharp as in the normal. Cell counts of reduced striate cortex in the enucleate show that area 17 is twice as thick as area 18, as is the case in the normal animal (Rockel et al. 1980). In the enucleate one of the numerous induced sulci can be seen lying posterior to the lunate sulcus. Arrowheads indicate the striate border. LS, lunate sulcus; Lat cal, lateral calcarine; IOS, inferior occipital sulcus; OT, occipital-temporal sulcus. Scale bar=2 mm. From Dehay et al. (1996 a)

cytoarchitecture, including the overall thickness in terms of numbers of neurons. We refer to this region as default striate cortex (DEC; Fig. 5). The reduced thickness of DEC compared to striate cortex in the early enucleates suggests either an increase in cell death or a decreased production of neurons following the enucleation. This has led us to hypothesise that the depletion of thalamic afferents was causing the reduction in the tangential extent of area 17 by inducing a major drop in the proliferation of area 17 precursors. This implies that thalamic fibers can influence the cycling behaviour of cortical precursors. The idea that afferents relaying information from the periphery can influence the proliferation of target cells had been substantiated in a series of pioneering experiments by Kollros, who showed that removal of the retina in the larval frog decreased the number of mitotic figures in the contralateral optic tectum (Kollros 1953, 1982). Similar findings have been reported in the insect eye (Selleck et al. 1992). We have

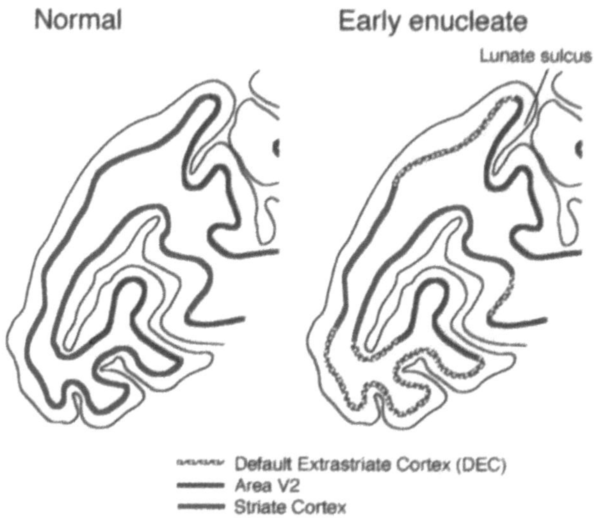

Fig. 5. Respecification of visual cortex. Effect of early enucleation (before embryonic day 81, E81) on the dimensions of striate cortex and the location of default extrastriate cortex (DEC). In the normal animal, striate cortex is surrounded by area V2 as shown in this parasagittal section. Similar sections in an early enucleate demonstrate that striate cortex is considerably reduced. Areal measurements show that the overall dimensions of the neocortex are not influenced by enucleation, so that the cortex that was destined to become striate cortex takes on a cytoarchitecture that is indistinguishable from area 18, and this region is referred to as DEC. Although we do not know the exact dimensions of DEC, we can assume that it extends as far as the limits of striate cortex in the normal animal. DEC can, in turn, be expected to be surrounded by presumptive area V2. From Dehay et al. (1996 b)

examined this issue *in vitro* and have been able to show that mouse cortical precursor cells that are cultured in thalamus conditioned medium show shorter Tc values and a decrease in the fraction of precursors that leave the cell-cycle (Dehay et al. 1995; Kennedy and Dehay 1995). These two effects on the cell-cycle radically increase the cycling fraction, leading to a rapid amplification of cell number. Similar effects are found with mitogenic neurotrophins, suggesting common mechanisms (Dehay et al. 1996 c).

What is the relevance of these findings to the *in vivo* condition? In the case of the mouse the regional differences are established prior to the arrival of thalamic afferents, so that these proliferative differences are intrinsic properties of the developing cortex. The *in vitro* findings show that cortical precursors have the ability to respond to diffusible factors that are released by quiescent thalamic neurons, so that regional differences of cell-cycle kinetics will be expected to be further modulated by thalamic afferents. However, once cortical precursors have elected to follow a particular pattern of proliferation, they would seem to show a large degree of autonomy, since cultured precursors of primate 17 and 18 continue to show very different cycling behaviour. Further, the characterisation of the cell-cycle in the culture experiments has enabled us to measure both the nature of the areal differences in the cell-cycle parameters as well as the phase of the cell cycle that is influenced by thalamic signals (Fig. 2). In both instances we found that the regulated parameter was the length of the G1 phase. This is particularly satisfying, firstly because the many control points in G1 are heavily implicated in control of the cycle in numerous systems (Korr 1978; Laskey et al. 1989; Pardee 1989; McAuley et al. 1993), and secondly, because modelling the parameters of cell-cycle of cortical precursors predicts that G1 would be the most likely regulated parameter (Caviness et al., this volume).

Emerging Patterns of Connectivity

Once cortical neurones have undergone their last round of mitosis in the germinal zones lining the ventricles, they migrate out to the cortical plate and differentiate in a fashion dictated at least partly by their birthdate.

Descending Pathways

However, within a given cortical layer there are numerous possible targets. For instance, although it is true that layer 5 neurons project to the mid-brain and spinal cord, different populations selectively target superior colliculus, pretectum, inferior olive and the pyramidal tract (Catsman-Berrevoets and Kuypers 1981; Keizer and Kuypers 1984; O'Leary and Stanfield 1985; Hallman et al. 1988; Killackey et al. 1989).

Numerous axonal tracing experiments claim that certain categories of neurones in immature cortex show a more widespread connectivity than in the adult. Hence, whereas projections to the spinal cord originate uniquely from layer 5 neurones in the rostral two-thirds of the adult rodent cortex, in the newborn of a variety of species the tangential distribution in the cortex is considerably larger (Stanfield et al. 1982; Bates and Killackey 1984; Joosten et al. 1987; Schreyer and Jones 1988 a, b; Meissirel et al. 1993). This finding has led to the suggestion that the functionally appropriate patterns of connectivity are pruned from an earlier uniform distribution and that the elimination of connections occurs in the late fetal period (Bates and Killackey 1984; O'Leary 1992). It is therefore hypothesised that axon elimination will be creating adult patterns of connectivity in the same way that cell death had been hypothesised to create the variations of neurone density underlying the cytoarchitectonics of the adult. The concept of emerging patterns through cell death was once again proposed to palliate a limited content in the genetic instruction (Bates and Killackey 1984; O'Leary 1992; O'Leary et al. 1992).

To investigate this issue we have compared the areal distribution of corticospinal neurones at different postnatal stages in the ferret (Meissirel et al. 1993). The ferret is a carnivore similar in many respects to the cat which presents the distinct advantage as a developmental model of being born at a much earlier developmental stage than either rats or cats. Hence the corticospinal pathway

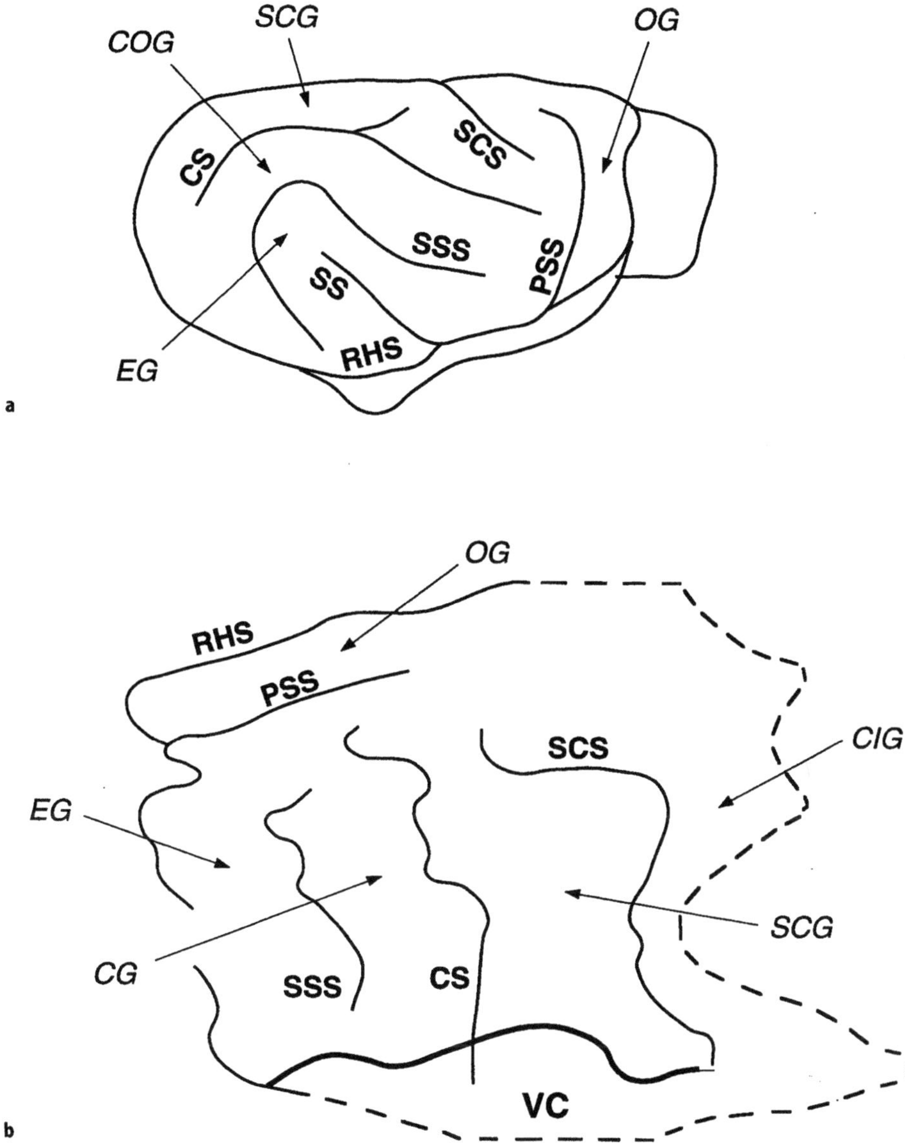

Fig. 6. Coronal sulci and gyri in the ferret. a. Lateral view of the brain. **b.** Flattened two-dimensional reconstruction of the cortex. CIG, cingulate gyrus; SCG, splenocruciate gyrus; COG, coronal gyrus; EG, ectosylvian gyrus; OG, orbital gyrus; SCS, spleniocruciate sulcus; CG, coronal gyrus; CS, coronolateral sulcus; SSS, suprasysylvian sulcus; SS, sylvian sulcus; PSS, presylvian sulcus; RHS, rhinal sulcus; VC, primary visual cortex. From Meissirel et al. (1993)

develops largely postnatally, which facilitates the experimental quantification of the areal distribution of corticospinal neurones at different developmental stages. This study shows, in the immature cortex, that although there is a more widespread distribution of corticospinal neurones, the distribution is non-uniform and importantly the posterior pole of the cortex containing the striate cortex is devoid of labeled neurones at all stages (Fig. 6, 7). Density measures of corticospinals showed that peak levels of labeling in the immature cortex had the same

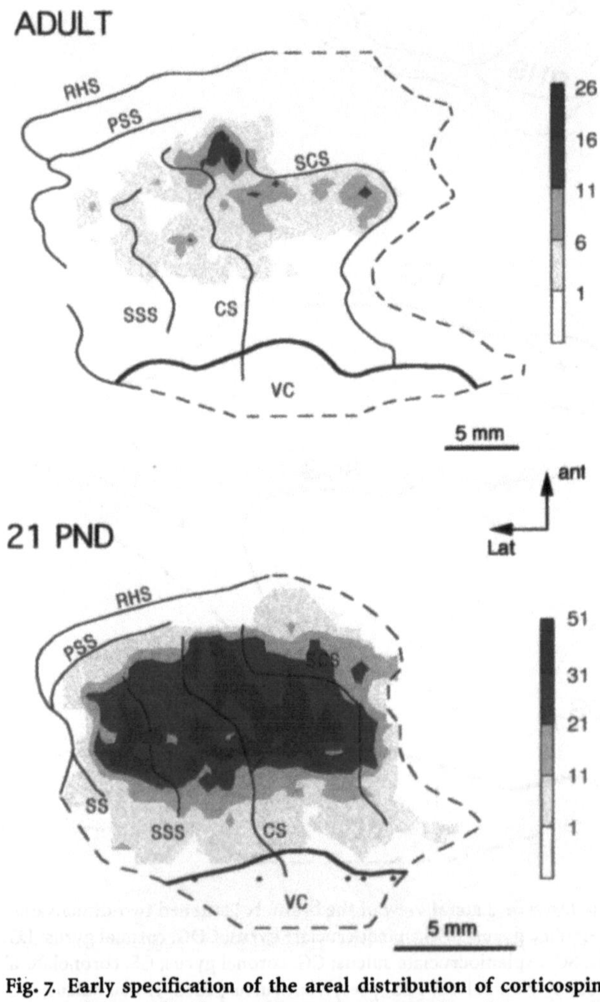

Fig. 7. Early specification of the areal distribution of corticospinal neurons. Injections of fast-blue were made immediately posterior to the decussation of the pyramidal tract in (**top**) adult and on postnatal day (PND) 3 pup (**bottom**) perfused on PND 21. Figure shows two-dimensional reconstructions of areal densities of labelled corticospinal neurons. Note that although labeled neurons stretch further posterior in the neonate, the distribution is highly non-uniform. Comparison with the young adult shows that peak levels in the neonate are located in those regions showing peak labeling in the adult. Abbreviations as in Figure 6

location in the sensorimotor cortex as in the adult. Comparison of the distribution profile of corticospinals in mature and adult cortex showed a very similar distribution, with higher absolute levels in the immature cortex. Further, there were cortical regions that were devoid of labeled corticospinals in the adult cortex and contained labeled neurones in the immature cortex but in these cases the level of labeling in the immature cortex was very low. This suggests that selective elimination played only a minor role in the maturation process and that it did not create the adult pattern but merely served to sharpen it. Subsequent studies of the areal densities of corticospinals in rodent have produced similar findings (Oudega et al. 1994) and support previous findings showing an early selective outgrowth of corticospinal projections from the sensorimotor cortex of the prenatal rodent (Schreyer and Jones 1988 a, b).

Corticocortical Pathways

Callosal connections. How do these findings in the corticospinal pathway compare to the development of corticocortical pathways? The concept of selective elimination of transient connections underlying the emergence of the adult pattern has been repeatedly suggested as being a major factor in the development of corticocortical connections. These studies in the 1970s and 1980s have to a large extent focused on connections between the cerebral hemispheres, where the restricted pattern found in the adult has been shown to be preceded by an immature widespread distribution. Hence in a number of species including rodents (Wise and Jones 1976; Ivy et al. 1979), lagomorphs (Chow et al. 1981), and cat (Innocenti et al. 1977), callosal projecting neurones are more widespread in the immature cortex so that, for instance, in the immature cortex, callosal neurones are found in area 17 of the immature cortex but not in the adult. However, in the monkey, area 17 is devoid of transient connections, suggesting that transient connectivity is not an obligatory step in the formation of connections (Dehay et al. 1988; Chalupa et al. 1989).

Despite the large number of studies in a variety of species, it remains difficult to judge the developmental significance of transient connections across the callosum. Is the adult pattern of callosal connectivity *created* from a perfectly uniform distribution by a selective elimination of the inappropriate connection or is the well-documented and massive reduction of connections perfectly random? If the latter were the case, then the mature distribution would prefigure in the immature distribution. To answer this question we need quantitative studies that compare the changing densities of callosal projecting neurones. Such studies are, alas, cruelly lacking. However, a recent study in rodent which provided a global visualisation of callosally projecting neurones suggested that there may be a nonuniform distribution of callosal projecting neurones in the immature cortex resembling that found in the adult cortex (Hernit et al. 1996).

Association Pathways

There have been altogether many fewer studies on association cortical pathways linking cortical areas within a single hemisphere. In many respects this is rather surprising, because it is the association connections that subserve the bulk of information processing in the cortex. An important functional consideration of cortical connections in the visual areas is the degree of convergence and divergence that will determine the magnification factor at ascending levels (Salin et al. 1989, 1992). Quantitative studies of the topography of intrinsic connections within cortical areas as well as extrinsic connections between areas show that, during development in the kitten (Kennedy et al. 1994) and monkey (Coogan and Van Essen 1996), values of divergence and convergence remain largely constant or show a moderate increase during the developmental process. These findings along with those in the thalamocortical pathway argue that the adult topography is the consequence of a progressive increase in the size and complexity of axonal arbors rather than the pruning back of an early exuberant projection (Friedlander and Martin 1989; Agmon et al. 1993; Kennedy et al. 1994; Catalano et al. 1996; Galuske and Singer 1996).

Functional Pathways

There are two major categories of cortical pathways: feedforward and feedback pathways (Fig. 8 and 9). These two categories need to be distinguished because recent results suggest that they follow very different developmental programs.

Feedforward pathways connect areas in an ascending series starting from the primary visual area which receive input from the thalamus (Maunsell and Van Essen 1983; Felleman and Van Essen 1991; Salin and Bullier 1995). Feedback projections connect in the opposite direction and go from higher to lower hierarchical levels. Anatomically these two pathways can be clearly distinguished. Feedforward projections are rostrally directed, originate from supragranular layers and

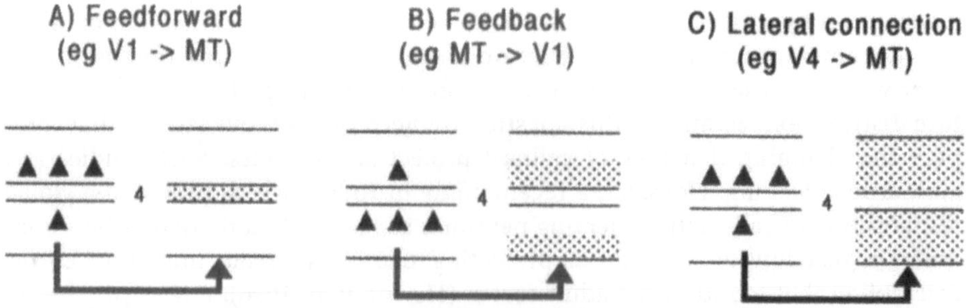

Fig. 8. Laminar distribution of cell bodies and terminals in cortical pathways according to direction of information flow. The same colour code is used as in Figure 9. From Kennedy et al. (1996 a). V1: Area V1 or area 17; MT: area V5 or STS; V4: area V4

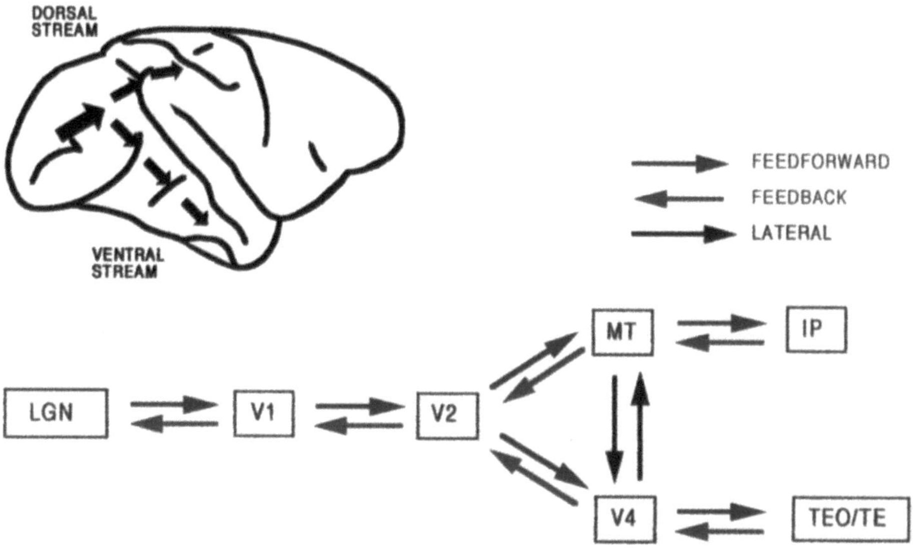

Fig. 9. Pathways in the visual cortex of the monkey. Colour codes as in Figure 8. Note MT is also referred to as V5. From Kennedy et al. (1996 a). V2: Area V2 or area 18; IP: inferior parietal area; TEO/TE: Temporal visual areas; LGN: Lateral geniculate nucleus

terminate in layer 4. Feedback projections are caudally directed, originate in infragranular layers and terminate outside of layer 4 (Kuypers et al. 1965; Cragg 1969; Spatz et al. 1970; Tigges et al. 1973; Lund et al. 1975; Martinez-Millan and Hollander 1975; Kaas and Lin 1977; Van Essen and Zeki 1978; Wong-Riley 1978, 1979; Rockland and Pandya 1979; Kennedy and Bullier 1985).

A characteristic feature of feedforward pathways in the visual system is their segregation into parallel pathways with distinct functional attributes (Ungerleider and Mishkin 1982; Bullier and Nowak 1995; Salin and Bullier 1995). Functional segregation of parallel pathways confronts the developing system with a rather unique problem that could be quite different from that of establishing point-to-point topographical connections linking sensory surfaces which we examined above. Early work in the kitten suggested that functional segregation might emerge from an early organisation that lacked segregation. In adult cat, injections of retrograde tracers in areas 18 and 19 reveal a patchy distribution of neurons in area 17 whereas similar injections in the immature cortex lead to a continuous distribution of labeled neurons (Bullier et al. 1984; Price and Blakemore 1985). However, these two columnar systems projecting to 18 and 19 are not truly segregated and, strictly speaking, might not correspond to two distinct functional pathways.

To test whether functional streams emerge from an immature situation where early-formed projections overlap, we have turned our attention to the ventral (temporal) and dorsal (parietal) directed pathways in monkey extrastriate cortex. In the monkey there are adjacent and interdigitizing territories in area

V2 which have different sets of inputs and in turn project to distinct targets. Regions in area V2 corresponding to the thin cytochrome oxidase bands and pale bands project to area V4 and temporal areas and are largely concerned with color and the spatial features of the stimulus. Regions in area V2 corresponding to the thick cytochrome oxidase bands project to area V5 or STS and parietal areas and are primarily concerned with motion analysis (DeYoe and Van Essen 1985; Livingstone and Hubel 1987; Shipp and Zeki 1989; Zeki and Shipp 1989; Lund et al. 1993; Nakamura et al. 1993; DeYoe et al. 1994; Levitt et al. 1995).

This system is particularly interesting for a developmental investigation since a strong expression of acetylcholinesterase demarcates the different area V2 territories during prenatal development: territories projecting to area V4 are densely labeled by acetylchlolinesterase (Barone et al. 1994). Injection of retrograde tracers in area V4 at the onset of pathway formation (some 60 days before birth) make it possible to examine the degree of segregation at different fetal ages (Barone et al. 1996). This experiment showed that, when the very first axons from area V2 arrive in the region of the white matter underlying area V4, they originate from the appropriate acetylcholine territories (Fig. 10). As development proceeds there is a rapid increase in the numbers of connections from the appropriate territories, so that when peak levels of connectivity are achieved 30 days before birth the segregation of the two pathways is complete (Fig. 10). The quantitative analysis of the distribution of retrograde labeling enabled us to show that it was the directed growth of the axons of neurones in the appropriate area V2 territories which establishes the segregation, and the late elimination of connections made only a very minor contribution to sharpening the early-formed pattern.

These findings in the development of the feedforward connections in the area V2 to V4 projection were surprising insofar as they showed that the cortico-cortical pathway develops a highly specific pattern very early in development, largely independently of late regressive events. The fact that the initial growth of the V2 axons is appropriately directed suggests that there are targeting mechanisms enabling the growing axons to recognise their appropriate trajectory. Such trophic mechanisms suggest that similar molecular cues are operant in cortex, as has been demonstrated in the ascending visual pathway of vertebrates as well as in insects (Goodman and Shatz 1993; Goodman 1994; Bolz et al. 1995, 1996; Miller et al. 1995; Henke-Fahle et al. 1996). These results would be compatible with

Fig. 10. Early specification of feedforward cortical pathways in the primate. Injection of retrograde tracers in area V4 reveals dense clusters of labeled neurones in area V2 (**a**). The adjacent section has been reacted for acetylcholinesterase (AChE) histochemistry (**b**). Counts of labeled neurones were made throughout all labeled and unlabeled AChE bands and define a labeling index (**c**). The labeling index provides a quantitative measure of the segregation of the V4 projection. These results show that the first V2 axons arrive in the vicinity of area V4 at around E108. At E112 when AChE bands can first be detected, labeled neurones already have a clustered distribution. In the following 15 days there is a massive increase in the percentages of labeled neurones in AChE bands and only a very modest increase in the interbands, indicating the role of directed growth in the clustered projection of V2 to V4. Arrow indicates birth. From Barone et al. (1996) ▶

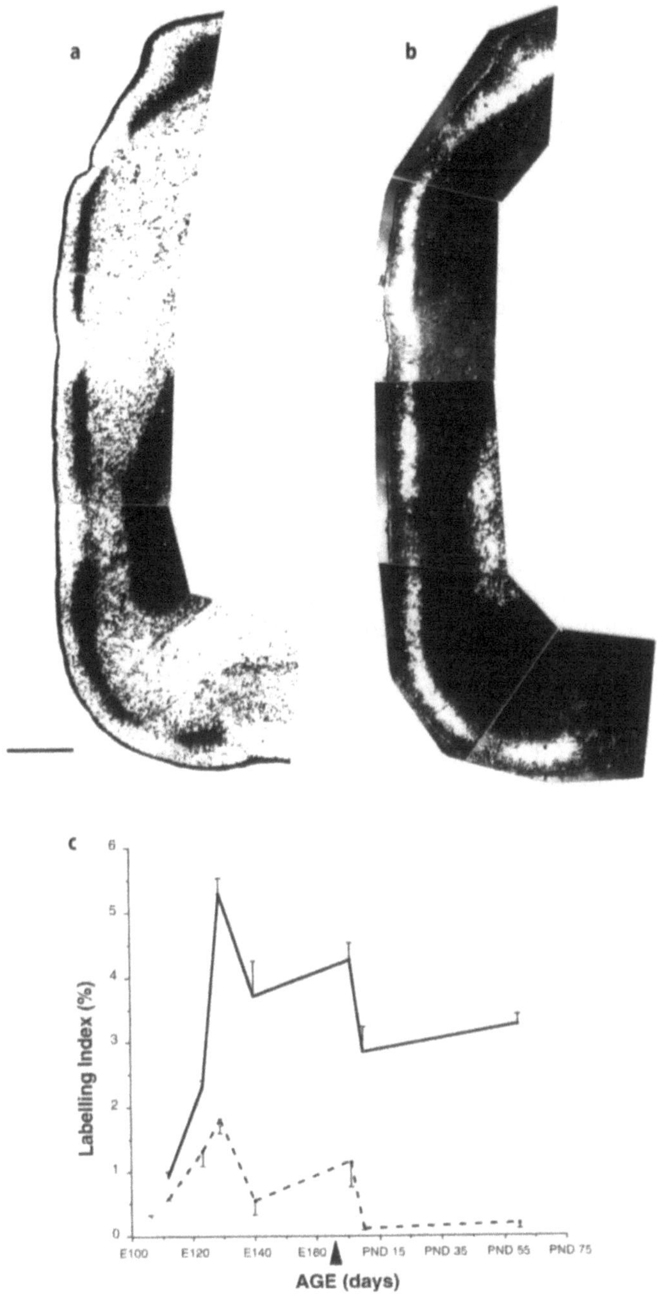
a
b
c
Labelling Index (%)
0
1
2
3
4
5
6
E100
E120
E140
E160
PND 15
PND 35
PND 55
PND 75
AGE (days)

the suggested role of molecular mechanisms at early stages of pathway formation and activity-dependent mechanisms in later stages of refinement (Katz and Shatz 1996; Tessier-Lavigne and Goodman 1996).

Feedback Pathways

The above results present a picture of the development of cortical connections following a model of early specification with little or no influence of late epigenetic events. However, observations of the development of the feedback pathways suggest that this only corresponds to half the story. Injections of retrograde tracers in monkey striate cortex show that, during development, connections are restricted to those cortical areas that are known to project to this area in the adult (Barone et al. 1995). However, in contrast to the invariance of the tangential pattern of connectivity, we found that there was massive change in the laminar distribution of retrogradely labeled neurones.

All cortical projections to area 17 are by definition caudal-directed and of the feedback configuration. Hence these projections arise predominantly from the infragranular layers. The strength of the input from the infragranular layers is variable, however, and is found to increase at distal sources of input (Kennedy and Bullier 1985). Hence, infragranular layers constitute nearly 95 % and supragranular layers 5 % of the area V5 input to striate cortex, whereas in area V2, which is adjacent to striate cortex, the infragranular contingent drops to 55 % and the supragranular layers provide 45 % of the input. Intermediate areas V3, V3A and V4 show intermediate and graded values. Injections made at the onset of pathway formation showed that the projection from all areas originate predominantly from the supragranular layers (Kennedy et al. 1989; Meissirel et al. 1993; Barone et al. 1995). As development proceeds into the postnatal period, the contribution from the infragranular layers progressively increases and the adult configuration is finally achieved a month after birth (Fig. 11). At all ages the afferent pattern of connectivity was observed with injections that were restricted to the cortical grey matter. By making paired, side-by-side injections of separate tracers we were able to further analyse the changing laminar distribution. This showed a decrease in supragranular layer projections at early stages of development and an increase in the proportion of infragranular layer projections at later stages (Barone et al. 1995).

These results are remarkable on several counts. Firstly, remodelling of the feedback pathways is a characteristic of primates. Similar analysis of labeling patterns in developing kittens shows weak remodelling (Batardière et al. 1996). Secondly, in contrast to the rapid development of the feedforward pathway, the remodelling of the feedback pathway occurs late in development. Thirdly, remodelling of feedback projections is a very general process found throughout a large expanse of neocortex. Injections made in area V4 throughout the developmental period show that feedback projections to this area arise principally from the supragranular layers and the major input from infragranular layers is not achieved

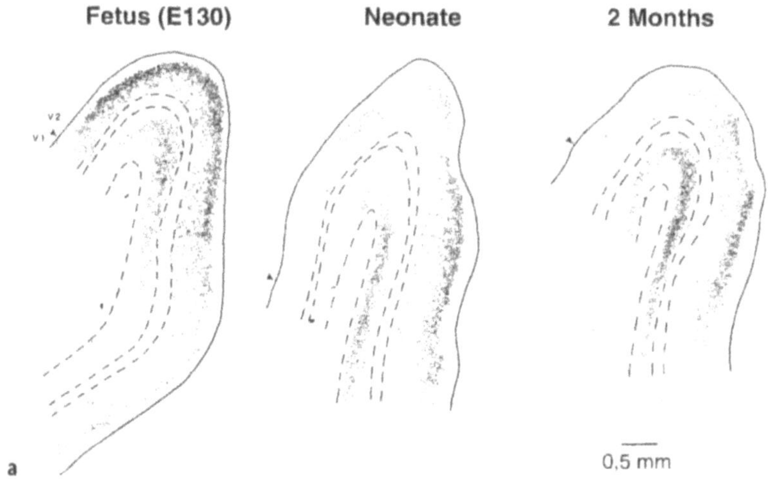

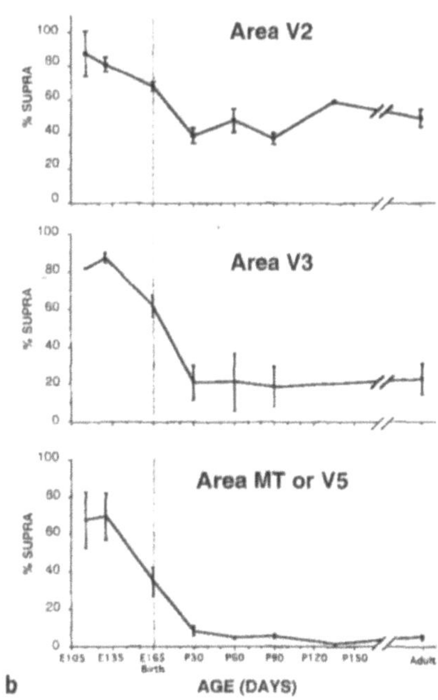

Fig. 11. Remodelling of feedback connections in the primate (a). Laminar distribution of labeled neurones in area V2 following injections in area V1. By computing the numbers of labeled neurones in supra- and infragranular layers throughout the projection zone it is possible to quantitatively describe the developmental reduction of supragranular layer neurones (**b**). These results show that remodelling of feedback pathways is particularly prolonged into late developmental stages of development. From Barone et al. (1995)

until after birth (Kennedy et al. 1996 a). These injections allow us to address the issue of whether remodelling is uniquely found in feedback projections. Those areas that provide a feedforward projection to area V4 (V2, V3 and V3A) were found to have projections that originate principally from the supragranular layers at all stages as in the adult.

We have recently addressed the issue of whether developmental remodelling of feedback pathways is a unique feature of the visual system. Preliminary findings show that this is not the case and suggest that similar processes are involved in the maturation of somatosensory pathways, where, as in the visual system, feedforward pathways are determined early and feedback pathways show a prolonged maturation.

The functional significance of the prolonged development of feedback pathways needs to be considered with respect to the physiological role of these pathways (Shao and Burkhalter 1996). Inactivation of a cortical area coupled with single unit recording in its target areas suggests that feedforward pathways provide the neural activity necessary for the elaboration of receptive field properties while feedback pathways provide modulatory influence of visual activation from outside the receptive field (James et al. 1995). Feedback projections are also implicated in the correlation of activity and are involved in the assembly-forming connections that may have important roles in perceptual grouping (Mignard and Malpeli 1991; Knierim and Van Essen 1992; Casagrande and Kaas 1994; Salin and Bullier 1995; Singer 1995; Miller 1996). Axons from infragranular and supragranular neurons might influence the microcircuitry of the target area in very different ways. Although we lack data in the monkey, there is evidence of a tendency in the cat and rat for supragranular layers to project more strongly to supragranular layers whereas infragranular layers project more equally to infragranular and supragranular layers (Henry et al. 1991; Coogan and Burkhalter 1993). Hence, the actual proportions of these projections could determine the physiological effect of the source area on the target area, and it could be expected that, in the more compartimentalised primate cortex, where direction of information flow is more clearly defined, the differences of targets of supra- and infragranular layers will likewise be more clearcut. Accordingly, the lower divergence and convergence values of supragranular layers might be related to their excitatory involvement in the elaboration of the receptive field response (Salin et al. 1992; Barone et al. 1995). Likewise, the high divergence values of the infragranular layers would make them more suited to their modulatory role. Although we lack in primates the anatomical evidence of the differences in the detailed anatomy of infra- and supragranular layer axon arbors, these considerations lead to several speculations. One possibility is that the supragranular projection in the immature primate has an excitatory function that is going to decrease enormously during development. This would mean that, throughout development, when supragranular projections in caudal-directed feedback pathways predominate, the flow of "excitatory" activity could be inverted with respect to the adult. The fact that this configuration is present during postnatal stages when sensory experience could be shaping neuronal response is of particular importance. Interestingly, it has recently been reported by Levitt and Lund (1996) that surround modulation was either weak or absent in area V1 of infant monkey.

Implication for Neurocognitive Development

An example of the sort of consequence that this could have is illustrated by recent findings on the functional neuroanatomy underlying language acquisition by comparing the cortical localization of language in children and adults (Thal et al. 1991; Holcomb et al. 1992; Mills et al. 1993; Neville 1995; Elman et al. 1996). This work shows that during development there is a progressive restriction of the cortical volume dealing with language, so that the cortical processing dealing with this function is considerably more distributed in the infant than in the adult. This could be subserved by just the sort of progressive restriction of connectivity that we find in the redistribution of projection neurones in feedback pathways.

The phenomenon of synaesthesia - the association of sound with certain colours in a small number of individuals - has been experimentally established in recent years (Baron-Cohen 1987; Baron-Cohen and Harrison 1993, 1996). Recent findings with PET support a neocortical origin (Paulesu et al. 1995). Conceptually, synaesthesia represents a breakdown in modularity which constitutes a definite challenge to our understanding of cognitive function. The present findings could relate to the breakdown of modularity in a number of ways. There could be variability in the extent of remodelling of feedback connections. Alternatively, the multimodal cortical functions set up in infancy could lead to permanent memory tracers in particular individuals (Kennedy et al. 1996 a).

A developmental theory of synaesthesia must attempt to take into account the finality of the mechanisms that lead to the adult condition. Bates has speculated on the possible advantage of a more extensive connectivity in the early stages of language acquisition (Bates 1994; Elman et al. 1996). It comes down to the infant's necessity to name objects and to bring about a convergence of the senses in an effort to facilitate that process. This could be a common feature in early stages of consciousness in humans and could require different levels of modularity that in turn require greater convergence at relatively low levels of cortical processing. Interestingly, our results on the remodelling of cortical connections suggest that, if this is a basis for early multimodal function, it is more pronounced in primates than in other orders and suggests prolonged and complex learning at the neocortical level. In conjunction with this, one would predict that the expression of NMDA receptor subtypes and the related extension of LTP should be prolonged in particular layers of the primate cortex (Fox and Daw 1993; Meskenaite et al. 1996).

These issues raise the more general question of the significance of plasticity in immature *and* adult cortex. Preattentive perceptual learning is a psychophysical response that reflects neural plasticity at early stages of cortical processing. Progressive areal restriction of cortical activity during the performance of a learned task has been demonstrated in human PET studies of preattentive perceptual learning in adults (Schiltz et al. 1995), which suggests an interesting parallel between the plasticity involved in preattentive perceptual learning and functional and anatomical restrictions found in development. One might expect other parallels between development and cortical function in the adult. For instance LTP-like mechanisms (Bliss and Lomo 1973; Artola and Singer 1987) could be involved in adult cortical plasticity (Kaas et al. 1990; Gilbert and Wiesel 1992) and preattentive perceptual learning (Fiorentini and Berardi 1980; Karni and Sagi 1991; Fahle et al. 1995; Schoups and Orban 1996) and be mediated by lateral or feedback connections (Eysel et al., personal communication). It is becoming increasingly clear that area V1 is not just the "retina of the cortex" and rather that is a node in a distributed system so that besides LTP, perceptual learning and functional plasticity, there is evidence of effects of attention on spatially processing (Motter 1993) as well as the signalling of "illusory" contours (Grosof et al. 1993). Developmental plasticity, could be more directly concerned with one or more of these attributes. The changes we report in the composition of the feedback pathways may influence function in the immature brain accordingly.

Conclusions

Elsewhere we have argued that cortical development is more usefully thought of in terms of the timing of specification events rather than as a balance between intrinsic (genetic) and extrinsic (epigenetic) control (Kennedy and Dehay 1993 a, b). Accordingly, the timing of the specification of a cortical feature will determine the extent to which environmental signals contribute to the determination of that feature. Comparison of the developmental timetable of different species leads to the notion of heterochrony (Gould 1977) and reveals selective adaptive strategies (Kennedy and Dehay 1993 b). In the present review we have provided evidence that the cardinal features of CNS development (cell proliferation and axonal growth and targeting) specify early in development highly precise areal patterns. Against this backdrop of early specification of neuron number and feedforward connections is the finding that regressive events sculpt out the connectivity of feedback projections. Of course, it may be that regressive events also contribute to the fine tuning of feedforward connections, possibly at the level of the microcircuitry. Nevertheless, the present results would suggest that the major differences in the physiological function of the immature cortex will concern feedback pathways more specifically.

Acknowledgments

Thanks to Jurgen Bolz and Jean Bullier for their helpful comments. Some of these speculations were discussed at an early stage of gestation during a seminar in the spring of 1996 at the Institute for Neuroinformatics, University of Zurich/ETH. This work was supported by Human Frontier grant RG-55/94B and BIOMED grant PL951604.

References

Agmon A, Yang LT, Odowd DK, Jones EG (1993) Organized growth of thalamocortical axons from the deep tier of terminations into layer-IV of developing mouse barrel cortex. J Neurosci 13:5365–5382
Angevine JBJ (1965) Time of neuron origin in the hippocampal region. An autoradiographic study in the mouse. Exp Neurol Suppl 2:1–70
Arimatsu Y (1994) Latexin: a molecular marker for regional specification in the neocortex. Neurosci Res 20:131–135
Arimatsu Y, Miyamoto M, Nihonmatsu I, Hirata K, Uratani Y, Hatanaka Y, Takiguchi-Hayashi K (1992) Early regional specification for a molecular neuronal phenotype in the rat neocortex. Proc Natl Acad Sci USA 89:8879–8883
Arimatsu Y, Nihonmatsu I, Hirata K, Takiguchi-Hayashi K (1994) Cogeneration of neurons with a unique molecular phenotype in layer-V and layer-VI of widespread lateral neocortical areas in the rat. J Neurosci 14:2020–2031
Artola A, Singer W (1987) Long-term potentiation and NMDA receptors in rat visual cortex. Nature 330:649–652
Barbe MF, Levitt P (1992) Attraction of specific thalamic input by cerebral grafts depends on the molecular identity of the implant. Proc Natl Acad Sci USA 89:3706–3710
Baron-Cohen S (1987) Hearing words and seeing color: an experimental investigation of a case of synesthesia. Perception 16:761–767
Baron-Cohen S, Harrison K (1993) Coloured speech perception: Is synaesthesia what happens when modularity breaks down? Perception 22:419–426
Baron-Cohen S, Harrison J (1996) Synaesthesia: Classic and contemporary readings. Blackwells, Oxford
Barone P, Dehay C, Berland M, Kennedy H (1994) Developmental changes in the distriubution of acetylcholinesterase in the extrastriate visual cortex of the monkey. Dev Brain Res 77:290–294
Barone P, Dehay C, Berland M, Bullier J, Kennedy H (1995) Developmental remodeling of primate visual cortical pathways. Cereb Cortex 5:22–38
Barone P, Dehay C, Berland M, Kennedy H (1996) Role of directed growth and target selection in the formation of cortical pathways: prenatal development of the projection of area V2 to area V4 in the monkey. J Comp Neurol 374:1–20
Batardière A, Barone P, Dehay C, Kennedy H (1996) Laminar remodelling of cortico-cortical projections in the cat. Soc Neurosci Abst 22:1016
Bates CA, Killackey HP (1984) The emergence of a discretely distributed pattern of corticospinal projection neurons. Dev Brain Res 13:265–273
Bates E (1994) Modularity, domain specificity and the development of language. Discuss Neurosci 10:136–149
Beaulieu C (1993) Numerical data on neocortical neurons in adult rat, with special reference to the GABA population. Brain Res 609:284–292
Blaschke AJ, Staley K, Chun J (1996) Widespread programmed cell death in proliferative and postmitotic regions of the fetal cerebral cortex. Development 122:1165–1174
Bliss TVP, Lomo P (1973) Long-lasting potentiation of synaptic transmission in the dendate area of the anaesthetized rabbit following stimulation of the perforant path. J Physiol 232:331–356
Bolz J, Kossel A, Bagnard D (1995) The specificity of interactions between the cortex and the thalamus ed. Development of the cerebral cortex. C Blakemore, John Wiley. 193:173–191
Bolz J, Castellani V, Mann F, Henke-Falhe S (1996) Specification of layer-specific connections in the developing cortex eds. Neural development and plasticity. R R Mize R S Erzurumlu eds. Amsterdam, Elsevier. 108:41–54
Brodmann K (1909) Vergleichende Lokalisationslehre der Grosshirnrinde in Ihren Prinzipien dargestellt auf Grund des Zellenbaus. Leipzig, Barth
Bullier J, Nowak LG (1995) Parallel versus serial processing. New vistas on the distributed organization of the visual system. Curr Opin Neurobiol 76:251–272

Bullier J, Dehay C, Kennedy H (1984) Axonal bifurcation and corticocortical connectivity in the kitten visual cortex. J Physiol 353:22P

Casagrande VA, Kaas JH (1994) The afferent, intrinsic, and efferent connections of primary visual-cortex in primates. Cereb Cortex 10:201–259

Catalano SM, Robertson RT, Killackey HP (1996) Individual axon morphology and thalamocortical topography in developing rat somatosensory cortex. J Comp Neurol 367:36–53

Catsman-Berrevoets CE, Kuypers HGJM (1981) A search for corticospinal collaterals to thalamus and mesencephalon by means of multiple retrograde fluorescent tracers in cat and rat. Brain Res 218:15–33

Caviness VS (1982) Neocortical histogenesis in the normal and reeler mice: a developmental study based upon [^{3}H]-thymidine autoradiography. Dev Brain Res 4:293–302

Caviness VS, Sidman RL (1973) Time of origin of corresponding cell classes in the cerebral cortex of normal and reeler mice: An autoradiographic analysis. J Comp Neurol 148:141–152

Chalupa L, Killackey HP, Snider CJ, Lia B (1989) Callosal projections neurons in area 17 of the fetal rhesus monkey. Dev Brain Res 46:303–308

Changeux JP (1983) L'homme neuronal. Fayard, Paris

Changeux JP, Danchin A (1976) Selective stabilisation of developing synapses as a mechanism for the specification of neural networks. Nature 264:705–712

Chow KL, Baumbach HD, Lawson R (1981) Callosal projections of the striate cortex in the neonatal rabbit. Exp Brain Res 42:122–126

Cohen-Tannoudji M, Babinet C, Wassef M (1994) Early determination of a mouse somatosensory cortex marker. Nature 368:460–463

Coogan TA, Burkhalter A (1993) Hierarchical organization of areas in rat visual-cortex. J Neurosci 13:3749–3772

Coogan TA, Van Essen DC (1996) Development of connections within and between areas V1 and V2 of macaque monkeys. J Comp Neurol 372:327–342

Cragg BG (1969) The topography of the afferent projections in the circumstriate visual cortex of the monkey studied by the Nauta method. Vision Res 9:733–747

Dehay C, Kennedy H, Bullier J, Berland M (1988) Absence of interhemispheric connections of area 17 during development in monkey. Nature 331:348–350

Dehay C, Horsburgh G, Berland M, Killackey H, Kennedy H (1989) Maturation and connectivity of the visual cortex in monkey is altered by prenatal removal of retinal input. Nature 337: 265–267

Dehay C, Horsburgh G, Berland M, Killackey H, Kennedy H (1991) The effects of bilateral enucleation in the primate fetus on the parcellation of visual cortex. Dev Brain Res 62:137–141

Dehay C, Giroud P, Berland M, Smart I, Kennedy H (1993) Modulation of the cell cycle contributes to the parcellation of the primate visual cortex. Nature 366:464–466

Dehay C, Guibert A, Savatier P, Cortay V, Kennedy H (1995) Modulation in vitro of the proliferation of rodent cortical precursors by a thalamic cytokine. Soc Neurosci Abstr 21:1507

Dehay C, Giroud P, Berland M, Killackey H, Kennedy H (1996 a) The contribution of thalamic input to the specification of cytoarchitectonic cortical fields in the primate: Effects of bilateral enucleation in the foetal monkey on the boundaries and dimensions of striate and extrastriate cortex. J Comp Neurol 367:70–89

Dehay C, Giroud P, Berland M, Killackey H, Kennedy H (1996 b) Phenotypic characterisation of respecified visual cortex subsequent to prenatal enucleation in the monkey: development of acetylcholinesterase and cytochrome oxydase patterns. J Comp Neurol 316:386–402

Dehay C, Cortay V, Savatier P, Kennedy H (1996 c) Quantitative analysis of the influence of neurotrophins on the cell cycle parameters of mouse cortical precursor cells. Soc Neurosci Abst 22:1478

DeYoe EA, Van Essen DC (1985) Segregation of efferent connections and receptive field properties in visual area V2 of the macaque. Nature 317:58–61

DeYoe EA, Felleman DJ, Van Essen DC, McClendon E (1994) Multiple processing streams in occipitotemporal visual cortex. Nature 371:151–154

Douglas RJ, Martin KAC, Whitteridge D (1988) Selective responses of visual cortical cells do not depend on shunting inhibition. Nature 332:642–644

Ebrahimi-Gaillard A, Roger M (1996) Development of spinal-cord projections from neocortical transplants heterotopically placed in the neocortex of newborn hosts is highly dependent on the embryonic locus of origin of the graft. J Comp Neurol 365:129–140
Ebrahimi-Gaillard A, Guitet J, Garnier C, Roger M (1994) Topographic distribution of efferent fibers originating from homotopic or heterotopic transplants: heterotopically transplanted neurons retain some of the developmental characteristics corresponding to their site of origin. Dev Brain Res 77:271–283
Elman JL, Bates EA, Johnson MH, Karmiloff-Smith A, Parisi D, Plunkett K (1996) Rethinking Innateness: A connectionist perspective on development. Cambridge, MA, MIT Press
Fahle M, Edelman S, Poggio T (1995) Fast perceptual-learning in hyperacuity. Vision Res 35:3003–3013
Felleman DJ, Van Essen DE (1991) Distributed hierarchical processing in the primate cerebral cortex. Cereb Cortex 1:1–47
Finlay BL, Pallas SL (1989) Control of cell number in the developing mammalian visual system. Prog Neurobiol 32:207–234
Finlay BL, Slattery M (1983) Local differences in amount of early cell death in neocortex predict adult local specializations. Science 219:1349–1351
Fiorentini A, Berardi N (1980) Perceptual learning specific for orientation and spatial frequency. Nature 287:43–44
Fox K, Daw NW (1993) Do NMDA receptor have a critical function in visual cortical plasticity? Trends Neurosci 16:116–122
Frantz GD, McConnell SK (1996) Restriction of late cerebral cortical progenitors to an upper-layer fate. Neuron 17:55–61
Friedlander MJ, Martin KAC (1989) Development of Y-axon innervation of cortical area 18 in the cat. J Physiol 416:183–213
Galuske RAW, Singer W (1996) The origin and topography of long-range intrinsic projections in cat visual cortex: a developmental study. Cereb Cortex 6:417–430
Gilbert CD, Wiesel TN (1992) Receptive-field dynamics in adult primary visual-cortex. Nature 356:150–152
Goodman CS (1994) The likeness of being - Phylogenetically conserved molecular mechanisms of growth cone guidance. Cell 78:353–356
Goodman CS, Shatz CJ (1993) Developmental mechanisms that generate precise patterns of neuronal connectivity. Cell 72:77–98
Gould SJ (1977) Ontogeny and phylogeny. Cambridge, MA, Harvard University Press
Grosof DH, Shapley RM, Hawken MJ (1993) Macaque-V1 neurons can signal illusory contours. Nature 365:550–552
Hallman LE, Schofield BR, Lin CS (1988). Dendritic morphology and axon collaterals of corticotectal, corticopontine, and callosal neurons in layer V of primary visual cortex of the hooded rat. J Comp Neurol 272: 149–160
Henke-Fahle S, Mann F, Gotz M, Wild K, Bolz J (1996) Dual action of a carbohydrate epitope on afferent and efferent axons in cortical development. J Neurosci 16:4195–4206
Henry GH, Salin PA, Bullier J (1991) Projections form area-18 and area-19 to cat striate cortex - divergence and laminar specificity. Eur J Neurosci 3:186–200
Hernit CS, Murphy KM, Van Sluyters RC (1996) Development of the visual callosal cell distribution in the rat: Mature features are present at birth. Visual Neurosci 13(5):923–943
Holcomb PJ, Coffey SA, Neville HJ (1992) Visual and auditory sentence processing: a developmental analysis using event-related brain potentials. Dev Neuropsychol 8:230–241
Honig LS, Herrmann K, Shatz CJ (1996) Developmental changes revealed by immunohistochemical markers in human cerebral cortex. Cerebral Cortex 6:794–806
Hubel DH, Wiesel TN (1977) Functional architecture of macaque monkey visual cortex. Proc R Soc Lond B 198:1–59
Innocenti GM, Fiore L, Caminiti R (1977) Exuberant projection into the corpus callosum from the visual cortex of newborn cats. Neurosci Lett 4:237–242

Ivy GO, Akers RM, Killackey HP (1979) Differential distribution of callosal projection neurons in the neonatal and adult rat: Brain Res 173:532–537

James AC, Hupé JM, Lomber SL, Payne B, Girard P, Bullier J (1995) Feedback connections contribute to center-surround interactions in neurons of monkey areas V1 and V2. Neurosci Abs 21:204

Joosten EAJ, Gribnau AAM, Dederen PJWC (1987) An anterograde tracer study of the developing corticospinal tract in the rat: three components. Dev Brain Res 36:121–130

Kaas JH, Lin CS (1977) Cortical projections in area 18 in owl monkeys. Vision Res 17:739–741

Kaas JH, Krubitzer LH, Chino Y, Longston AL, Polley EH, Blair N (1990) Reorganization of retinotopic cortical preps in adult mammals after lesions of the retina. Science 248:229–231

Karni A, Sagi D (1991) Where practice makes perfect in texture discrimination: Evidence for primary visual cortex plasticity. Proc Natl Acad Sci USA 88:4966–4970

Katz LC, Shatz CJ (1996) Synaptic activity and the construction of cortical circuits. Science 274:1133–1138

Keizer K, Kuypers HGJM (1984) Distribution of corticospinal neurons with collaterals to lower brain stem reticular formation in cat. Exp Brain Res 54:107–120

Kennedy H, Bullier J (1985) A double-labelling investigation of the afferent connectivity to cortical areas V1 and V2 of the macaque monkey. J Neurosci 5:2815–2830

Kennedy H, Dehay C (1993 a) Cortical specification of mice and men. Cereb Cortex 3:27–35

Kennedy H, Dehay C (1993 b) The importance of developmental timing in cortical specification. Perspect Dev Neurobio 1: 93–99

Kennedy H, Dehay C (1993 c) The relevance of primate corticogenesis for understanding the emergence of cognitive abilities in man. In: Developmental neurocognition: Speech and face processing in the first year of life. B De Boysson-Bardies; S de Schonen, P Jusczyk, P McNeilage and J. Morton eds., Kluwer Academic Publishers. 69:17–30

Kennedy H, Bullier J, Dehay C (1989) Transient projections from STS to area 17 in the newborn monkey. Proc Natl Acad Sci USA 86:8093–8097

Kennedy H, Dehay C (1995) In vivo and in vitro evidence for areal commitment of neuron precursors in the ventricular zone of the neocortex. In: Bock GR, Cardew G (eds) Development of the cerebral cortex, Ciba Foundation Symposium 193 Chichester, John Wiley, pp 251–257

Kennedy H, Salin P, Bullier J, Horsburgh G (1994) Topography of developing thalamic and cortical pathways in the visual-system of the cat. J Comp Neurol 348:298–319

Kennedy H, Batardiere A, Dehay C, Barone P (1996 a) Synaesthesia: implications for developmental neurobiology. In: Baron-Cohen S, Harrison J (eds) Synaesthesia: classic and contemporary readings. Blackwells, Oxford, pp 243–256

Kennedy H, Dehay C, Ziegel G, Smart I, Giroud P, Berland M (1996 b) Birthdating experiments in embryonic monkey reveals distinct patterns of cell proliferation and migration in presumptive visual areas 17 and 18. Soc Neurosci Abst 22:525

Killackey H (1990) Neocortical expansion: an attempt toward relating phylogeny and ontogeny. J Cogn Neurosci 2:1–17

Killackey HP, Koralek KA, Chiaia NL, Rhoades RW (1989) Laminar and areal differences in the origin of the subcortical projection neurons of the rat somatosensory cortex. J Comp Neurol 282:428–445

Knierim JJ, Van Essen DC (1992) Neuronal responses to static texture patterns in area V1 of the alert macaque monkey. J Neurophysiol 67:961–980

Kollros JJ (1953) The development of the optic lobes in the frog: I The effects of unilateral enucleation in embryonic stages. J Exp Zool 123:153–187

Kollros JJ (1982) Peripheral control of midbrain mitotic activity in the frog. J Comp Neurol 205:171–178

Korr H (1978) Proliferation of different cell types in the brain. Berlin.

Kuypers HGJM, Szwarcbart MK, Mishkin M, Rosvold HE (1965) Occipitotemporal corticortical connections in the rhesus monkey. Exp Neurol 11:245–262

Laskey RA, Fairman MP, Blow JJ (1989) S Phase of the cell cycle. Science 246:609–613

Levitt JB, Lund JS (1996) Asynchronous development of receptive field properties and clustered horizontal connections in macaque striate cortex. Soc Neurosci Abst 22:491
Levitt JB, Yoshioka T, Lund JS (1995) Connections between the pulvinar complex and cytochrome oxidase-defined compartments in visual area V2 of macaque monkey. Exp Brain Res 104:419–430
Livingstone MS, Hubel DH (1987) Connections between layer 4B of area 17 and the thick cytochrome oxidase rich stripes of area 18 in the squirrel monkey. J Neurosci 7:3371–3377
Lund JS, Lund RD, Hendrickson AE, Bunt AH, Fuchs AF (1975) The origin of efferent pathways from the primary visual cortex of the macaque monkey as shown by retrograde transport of horseradish peroxydase. J Comp Neurol 164:287–304
Lund JS, Yoshioka T, Levitt JB (1993) Comparison of intrinsic connectivity in different areas of macaque monkey cerebral cortex. Cereb Cortex 3:148–162
Martinez-Millan L, Hollander H (1975) Cortico-cortical projections from striate cortex of the squirrel monkey (saimiri sciureus). A radioautographic study. Brain Res 83:405–417
Maunsell JHR, Van Essen DC (1983) The connections of the middle temporal visual area (MT) and their relationship to a cortical hierarchy in the macaque monkey. J Neurosci 3:2563–2586
McAuley A, Werb Z, Mirkes PE (1993) Characterization of unusually rapid cell cycles during rat gastrulation. Development 117:873–883
McConnell SK (1988) Fates of visual cortical neurons in the ferret after isochronic and heterochronic transplantation. J Neurosci 8:945–974
McConnell SK (1991) The generation of neuronal diversity in the central-nervous-system. Annu Rev Neurosci 14:269–300
McConnell SK, Kaznowski CE (1991) Cell-cycle dependence of laminar determination in developing neocortex. Science 254:282–285
Meissirel C, Dehay C, Kennedy H (1993) Transient cortical pathways in the pyramidal tract of the neonatal ferret. J Comp Neurol 338:193–213
Meskenaite V, Kennedy H, Dehay C, Martin KAC (1996) Postnatal maturation of AMPA and NMDA-type glutamate receptors in macaque monkey striate cortex. Soc Neurosci Abst 22:542
Mignard M, Malpeli JG (1991) Paths of information-flow through visual-cortex. Science 251:1249–1251
Miller B, Sheppard AM, Bicknese AR, Pearlman AL (1995) Chondroitin sulfate proteoglycans in the developing cerebral cortex: The distribution of neurocan distinguishes forming afferent and efferent axonal pathways. J Comp Neurol 355:615–628
Miller MW, Kuhn PE (1995) Cell cycle kinetics in fetal rat cerebral cortex: effects of prenatal treatment with ethanol assessed by a cumulative labeling technique with flow cytometry. Alcohol Clin Exp Res 19:233–237
Miller R (1996) Neural assemblies and laminar interactions in the cerebral cortex. Biol Cybern 75:253–251
Mills DL, Coffey-Corina SA, Neville HJ (1993) Language acquisition and cerebral specialization in 20-month-old infants. J Cogn Neurosci 5:317–334
Moraillon B, Polleux F, Dehay C, Kennedy H (1996) Dynamic modelisation of cortical neuron production. Soc Neurosci Abst 22:525
Motter BC (1993) Focal attention produces spatially selective processing in visual cortical areas V1, V2, and V4 in the presence of competing stimuli. J Neurophysiol 70:909–919
Mountcastle VB (1957) Modality and topographic properties of single neurons of cat's somatic sensory cortex. J Neurophysiol 20:408–434
Nakamura H, Gattass R, Desimone R, Ungerleider LG (1993) The modular organization of projections from area-V1 and area-V2 to area-V4 and TEO in macaques. J Neurosci 13(9):3681–3691
Neville HJ (1995) Developmental specificity in neurocognitive development in humans. In: Gazzaniga MS (ed) The cognitive neurosciences. Cambridge, MA, MIT Press, pp 219–231
O'Leary DDM (1989) Do cortical areas emerge from a protocortex? Trends Neurosci 12:400–406
O'Leary DDM (1992) Development of connectional diversity and specificity in the mammalian brain by the pruning of collateral projections. Curr Opin Neurobiol 2:70–77

O'Leary DDM, Stanfield BB (1985) Occipital cortical neurons with transient pyramidal tract axons extend and maintain collaterals to subcortical but not intracortical targets. Brain Res 336:326–333
O'Leary DDM, Schlaggar BL, Stanfield BB (1992) The specification of sensory cortex – lessons from cortical transplantation. Exp Neurol 115:121–126
Oppenheim RW, Cole T, Prevette D (1989) Early regional variations in motoneuron numbers arise by differential proliferation in the chick embryo spinal cord. Dev Biol 133:468–474
Oudega M, Varon S, Hagg T (1994) Distribution of corticospinal motor-neurons in the postnatal rat – quantitative evidence for massive collateral elimination and modest cell-death. J Comp Neurol 347:115–126
Pardee AB (1989) G1 events and regulation of cell proliferation. Science 246:603–608
Paulesu E, Harrison J, Baron-Cohen S, Watson JDG, Goldstein L, Heather J, Frackowiak RSJ (1995) The anatomy of coloured hearing. A positron tomography activation study of colour-word. Brain 118:661–676
Payne BR, Lomber SG, Villa AE, Bullier J (1996) Reversible deactivation of cerebral network components. Trends Neurosci, 19:535–542
Polleux F, Dehay C, Kennedy H (1995) Study of early regionalisation of the ventricular zone and its role in the production of distinct neocortical areas. Soc Neurosci Abstr 21:1513
Polleux F, Dehay C, Kennedy H (1997) The timetable of laminar neurogenesis contributes to the specification of cortical areas in mouse isocortex. J Comp Neurol, 385:95–116
Price DJ, Blakemore C (1985) Regressive events in the postnatal development of association projections in the visual cortex. Nature 316:721–724
Purves D, Lichtman JW (1985) Principles of neural development. Sunderland, Sinauer
Rakic P (1974) Neurons in rhesus monkey visual cortex: systematic relation between time of origin and eventual disposition. Science 183:425–427
Rakic P (1976) Prenatal genesis of connections subserving ocular dominance in the rhesus monkey. Nature 261:467–471
Rakic P (1977) Genesis of the dorsal lateral geniculate nucleus in the rhesus monkey: site and time of origin, kinetics of proliferation, routes of migration and pattern of distribution of neurons. J Comp Neurol 176:23–52
Rakic P (1982) Early developmental events: cell lineages, acquisition of neuronal positions and areal and laminar development. Neurosci Res Prog Bull 20:439–451
Rakic P (1988) Specification of cerebral cortical areas. Science 241:170–176
Rakic O (1995) A small step for the cell, a giant leap for mankind – a hypothesis of neocortical expansion during evolution. Trends Neurosci 18:383–388
Rakic P, Nowakowski RS (1981) The time and site of origin of neurons in the hippocampal region of the rhesus monkey. J Comp Neurol 196:99–128
Rakic P, Suner I, Williams RW (1991) A novel cytoarchitectonic area induced experimentally within the primate visual-cortex. Proc Natl Acad Sci USA 88:2083–2087
Rhoades RW, Wall JT, Chiaia NL, Bennett-Clarke CA, Killackey HP (1993) Anatomical and functional-changes in the organization of the cuneate nucleus of adult-rats after fetal forelimb amputation. J Neurosci 13:1106–1119
Rockel AJ, Hiorns RW, Powell TPS (1980) The basic uniformity in structure of the neocortex. Brain 103:221–244
Rockland KS, Pandya DN (1979) Laminar origins and terminations of cortical connections to the occipital lobe in the rhesus monkey. Brain Res 179:3–20
Salin PA, Bullier J (1995) Corticocortical connections in the visual system: structure and function. Physiol Rev 75:107–154
Salin PA, Bullier J, Kennedy H (1989) Convergence and divergence in the afferent projections to cat area 17. J Comp Neurol 283:486–512
Salin PA, Girard P, Kennedy H, Bullier J (1992) Visuotopic organization of corticocortical connections in the visual system of the cat. J Comp Neurol 320:415–434
Savatier P, Dehay C, Dumas AM; Cortay V, Kennedy H (1995) Facs analysis of cell parameters in vitro of monkey cortical precursors. Soc Neurosci Abstr 21:44

Schiltz C, Dubois S, Bodart JM, Michel C, Orban G, Roucoux A (1995) Visual orientation discrimination: effect of training on brain activity, a PET study. Soc Neurosci Abst 21:1423
Schlaggar BL, O'Leary DDM (1991) Potential of visual-cortex to develop an array of functional units unique to somatosensory cortex. Science 252:1556–1560
Schmahl W (1983) Developmental gradient of cell cycle in the telencephalic roof of the fetal NMRI-mouse. Anat Embryol 167:355–364
Schoups AA, Orban GA (1996) Interocular transfer in perceptual learning of a pop-out Discrimination task. Proc Natl Acad Sci USA 93:7358–7362
Schreyer DJ, Jones EG (1988 a) Axon elimination in the developing corticospinal tract of the rat. Dev Brain Res 38:103–119
Schreyer DJ, Jones EG (1988 b) Topographic sequence of outgrowth of corticospinal axons in the rat: a study using retrograde axonal labeling with fast blue. Dev Brain Res 38:89–101
Schultze B, Korr H (1981) Cell kinetic studies of different cell types in the developing and adult brain of the rat and the mouse: a review. Cell Tissue Kinet 14:309–325
Schultze B, Nowak B, Maurer W (1974) Cycle times of the neural epithelial cells of various types of neuron in the rat. An autoradiographic study. J Comp Neurol 158:207–218
Selleck SB, Gonzalez C, Glover DM, White K (1992) Regulation of the G1-S transition in postembryonic neuronal precursors by axon ingrowth. Nature 355:253–255
Shao Z, Burkhalter A (1996) Different balance of excitation and inhibition in forward and feedback circuits of rat visual cortex. J Neurosci 16:7353–7365
Shatz CJ (1992) Dividing up the neocortex. Science 258:237–238
Shipp S, Zeki S (1989) The organization of connections between areas V5 and V2 in macaque monkey visual cortex. Eur J Neurosci 1:308–331
Singer W (1995) Development and plasticity of cortical processing architectures. Science 270:758–764
Skoglund TS, Pasher R, Berthold CH (1996) Heterogeneity in the columnar number of neurons in different neocortical areas in the rat. Neurosci Lett 208:97–100
Spatz WB, Tigges J, Tigges M (1970) Subcortical projections, cortical associations, and some intrinsic interlaminar connections of the striate cortex in the squirrel monkey (saimiri). J Comp Neurol 140:155–174
Stanfield BB, O'Leary DDM, Fricks C (1982) Selective collateral elimination in early postnatal development restricts cortical distribution of rat pyramidal tracts neurons. Nature 298:371–373
Takahashi T, Nowakowski RS, Caviness VS (1995) The cell-cycle of the pseudostratified ventricular epithelium of the embryonic murine cerebral wall. J Neurosci 15:6046–6057
Takahashi T, Nowakowski RS, Caviness VS (1996) The leaving or Q fraction of the murine cerebral proliferative epithelium: A general model of neocortical neurogenesis. J Neurosci 16:6183–6196
Tessier-Lavigne M, Goodman CS (1996) The molecular biology of axon guidance. Science 274:1123–1133
Thal DJ, Marchman V, Stiles J, Aram D, Trauner D, Nass R, Bates E (1991) Early lexical development in children with focal brain injury. Brain Lang 40:491–527
Tigges J, Spatz WB, Tigges M (1973) Reciprocal point-to-point connections between parastriate and striate cortex in the squirrel monkey (saimiri). J Comp Neurol 148:481–490
Ungerleider LG, Mishkin M (1982) Two cortical visual systems. In: Ingle DJ, Goodale MA, Mansfield RJR (eds) Analysis of visual behavior. Cambridge, Massachusetts, MIT Press, pp 549–586
Van Essen DC, Zeki S (1978) The topographic organization of rhesus monkey prestriate cortex. J Physiol 277:193–226
Waechter RV, Jaensch B (1972) Generation time of the matrix cells during embryonic brain development: an autoradiographic study in rats. Brain Res 46: 235–250
Wise SP, Jones EJ (1976) The organization and postnatal development of the commisural projection of the rat somatosensory cortex. J Comp Neurol 168:313–344
Wong-Riley M (1978) Reciprocal connections between striate and prestriate cortex in the squirrel monkey as demonstrated by combined peroxydase histochemistry and autoradiography. Brain Res 147:159–164
Wong-Riley M (1979) Changes in the visual system of monocularly sutured or enucleated cats demonstrable with cytochrome oxidase histochemistry. Brain Res 171:11–28

Zeki SM, Shipp S (1989) Modular organization between areas V2 and V4 of macaque monkey visual cortex. Eur J Neurosci 1:494–506

Zeki SM (1978) Functional specialisation in the visual cortex of the rhesus monkey. Nature 274:423–428

Mechanisms of Neuronal Migration

C. Zheng, E. Ross, J. Alder, M. Hofer, J. Millonig, K. Millen, P. Kofuji, N. Davidson, H. Lester, N. Heintz, and M.E. Hatten

Abstract

The migration of neuronal precursors from germinal zones to neuronal layers is critical to the establishment of functional neural circuitry in developing brain. Previous work has identified a protein, astrotactin (ASTN), as the neuronal ligand for locomotion along the radial glial fiber system. Molecular cloning of astrotactin reveals that the molecule is a membrane associated glycoprotein with three EGF motifs and two FNIII domains. As shown by in vitro studies, the expressed bacterial fusion protein functions in neuronal migration, suggesting that astrotactin provides a mechanism for glia-guided migration. Genetic analyses indicate that the gene is located on Chromosome 1 in the mouse, in a region syntenic with 1q25–27 in humans. Mutations in this area of the human genome are associated with microcephaly and other developmental anomalies.

The mutant mouse weaver, an animal with inherited defects in migration, carries a single base pair change in the GIRK2 potassium channel. Neurophysiological studies have shown that the mutant channel becomes nonselective and that this defect leads to the weaver phenotype. Thus defects in molecular mechanisms of cell locomotion and in the assembly and function of ion channels are likely to play a major role in generating developmental deficits that underlie brain malformations.

Introduction

One of the most remarkable features of the developing nervous system is the wide-ranging migrations by precursor cells. These movements are orchestrated to promote the differentiation of an astonishing array of phenotypes and to arrange young neurons into the most complex biological structure known, the vertebrate brain (Ramon y Cajal 1911). In man, more than a billion cells, including hundreds of different types of neurons, utilize directed cell migrations to assemble into neural networks that comprise some 1011 synaptic contacts. In vertebrate brain, this complexity of cellular structure and synaptic connections is achieved by the construction of novel cellular organization, most notably the laminar arrangement of neurons in higher cortical regions. Thus, central questions in the nervous system are: why is cellular migration pervasive in the forma-

A. M. Galaburda / Y. Christen (Eds.)
Normal and Abnormal Development of Cortex
© Springer-Verlag Berlin Heidelberg New York 1997

tion of these structures, and how is the movement of cells across long distances achieved? In this chapter, we shall focus on the role of directed cell migrations in the execution of specific programs of neuronal differentiation.

As first documented in studies on human fetal development, and comprehensively reviewed by Sidman and Rakic (1973), the movements of precursor cells away from CNS germinal zones lead to the formation of three general classes of structures: layered structures with predominantly radial migration patterns, including the cortex, the hippocampal formation and the cerebellar cortex; layered structures formed by mixed radial and tangential migration, including the retina and spinal cord; and non-layered structures, including the brainstem, mesencephalon and diencephalon, where free-ranging movements of multipotent precursor cells result in the formation of nuclear structures.

Formation of the Basic Embryonic Zones

As rapid cell division thickens the proliferative zone, the emerging cortical structure progresses from a simple neuroepithelial sheet into a complex multi-laminar system. The first step in this process involves the creation of the marginal zone (M), a cell-sparse zone from which nuclei are apparently excluded during interkinetic movements in the epithelium. Second, a zone of ingrowing, afferent axons appears, forming an intermediate zone (I) between the ZV and the marginal zone. In the cortex, these axons are thought to comprise pioneer afferent axons extending between areal divisions (Schaper 1897; His and Ruffett 1899; Sauer 1935). The neuroepithelium thus progresses from a single layer of dividing cells to three and then four zones, termed the ventricular zone (proliferating cells), intermediate zone (axons), pre-plate (postmitotic neuronal precursors) and marginal zone (Boulder Committee 1970).

The importance of these rudimentary embryonic layers as building blocks for the subsequent laminar architecture of the cortex is illustrated in cases of cortical malformations arising from neurogenetic disorders. In such cases, disturbances in these apparently simple early events of cortical development reverberate spatially through a wide area. Although we shall focus on the process of migration, because relatively more information is available, little is known about the mechanisms that halt the movement of cells and form layers. Surprisingly, the discovery of the first functional role for an extracellular matrix molecule in establishing cortical laminae has come, not from localization of candidate molecules, but from neurogenetic studies on the reeler neurological mutant mouse, an animal with inherited defects in the formation of the embryonic layers (Curran 1994; Rakic and Caviness 1995).

The Cerebellar Granule Cell Presents a Model System for CNS Cell Development

The cerebellar granule neuron precursor cell population presents a model for CNS neuronal migration. As first described by Ramon y Cajal (1889), precursors of the granule cell undergo a highly unusual developmental scheme. Proliferating at the edge of the neuroepithelium (rhombic lip), where there is no overlying cellular structure, this pool of dividing precursor cells streams rostrally across the lip and onto the surface of the anlage (Miale and Sidman 1961), where they establish a displaced germinal zone, the external germinal layer (EGL); (Hatten and Heintz 1995).

Precursor cell proliferation continues within the EGL until about P15, when the zone disappears due to the inward migration of postmitotic cells (Rakic 1972; Roman y Cajal 1911) to form the internal granule cell layer (IGL). Whereas the target neurons of the granule cell undergo a short period of proliferation within the VZ of the cerebellar anlage, from approximately E11–E14 in the mouse (Miale and Sidman 1961; Altman and Bayer 1985), the prolonged period of clonal expansion of EGL precursor cells within the secondary germinal zone, from E13 until P15, generates a huge population of cells. In adulthood, granule neurons outnumber Purkinje cells by 250:1 in the mouse, a figure that rises to approximately 400:1 in humans.

As first realized by Ramon y Cajal (1889, 1911), the granule cell represents one of the most amenable classes of CNS neurons for developmental studies (Hatten and Heintz 1995). As cells in the EGL proliferate, some cells begin to exit the cycle and undergo differentiation, as evidenced by the formation of axons, the parallel fibers. Molecular biological studies indicate that the subdivision of the EGL into a superficial zone of dividing cells and an inner zone of differentiating cells is accompanied by the expression of genes that are differentially regulated at the transcriptional level during cerebellar development (Kuhar et al. 1993). In situ localization of these genes reveals that granule neuron differentiation can be divided into at least four discrete stages: neurogenesis, initiation of neuronal differentiation, axon outgrowth and migration, and formation of synaptic connections (Hatten and Heintz 1995).

To provide a genetic analysis of the molecular signals that induce the early steps of granule neuron differentiation, including cell migration, extensive studies have been performed with neurological mutant mice (Heintz et al. 1993). In the neurological mouse mutant weaver, CNS precursor cells in the EGL of the cerebellar cortex proliferate normally (Rezai and Yoon 1972), but fail to undergo early steps in differentiation: they neither extend long neurites nor migrate along glial fibers (Rakic and Sidman 1973; Sotelo and Changeux 1974). Experimental analyses of the site of action of the weaver gene – by cell mixing experiments in vitro (Hatten et al. 1986) and by the production of chimeras (Goldowitz and Mullen 1982; Goldowitz 1989) – demonstrated that the weaver gene acts in granule cells. Gao et al. (1992) found that the weaver granule cell phenotype was rescued by local cell interactions with wild type granule cells in vitro and in vivo.

This finding suggested that differentiation of immature EGL cells required local cell-cell interactions among neighboring granule cells.

Surprisingly, in the weaver strain a point mutation was identified in the mouse GIRK2 gene, which encodes a G protein-gated inwardly rectifying K+ channel (Patil et al. 1995). GIRK channels [HAL1] affect the excitability of cardiac atrial myocytes and of several types of neurons and are activated by direct interaction with G proteins (Pfaffinger et al. 1985; North 1989, Clapham 1994; Kurachi 1995). To date, five members of the GIRK gene family (GIRK1 through GIRK5) have been isolated. It is presently thought that GIRK channels are heteromultimers of two distinct GIRKs (Krapivinsky et al. 1995; Kofuji et al. 1995; Yang et al. 1995). The mutation found in weaver GIRK2 (wvGIRK2) exchanges a glycine residue for a serine in the main pore-forming domain (Patil et al. 1995). This amino acid is consreved in nearly all inwardly rectifying and voltage-gated K+ channels (Heginbotham et al. 1994; Doupnik et al. 1995).

After the GIRK channel was proposed to be a candidate for the weaver gene by Cox and his colleagues, we studied the expression of GIRK channels in cerebellum, the electrophysiology of weaver granule neurons, and the functional properties of the weaver GIRK2 mutation in a heterologous expression system (Kofuji et al. 1996). We found mRNA for GIRK2, as well as for GIRK1 and GIRK3, expressed in cerebellar granule cells and elsewhere in the CNS during development. We also detected GIRK currents in cultured +/+ P6 granule cells, but not in P6 wv/wv granule cells. On the other hand, the wv granule cells acquired an additional Na+ conductance. These properties of the wv granule neurons are explained by functional analysis of the mutant channel in Xenopus oocytes: it is both constitutively active and permeable to Na+. Thus weaver presents a case of failed migration owing to mutations in ion channels. The details of how this channel function impacts on migration remain to be demonstrated, as the defects in conductance of the channel do not damage cells in early phases of development, i.e., during proliferation, or cells located in the lateral aspect of the cerebellar cortex.

Further cues on early steps in cell differentiation, steps required for the cells to acquire features needed for migration, will likely come from studies on the GIRK2 channel. In particular, the control of the Na+ levels inside the cell and the role of G protein signaling systems will be interesting avenues to pursue. The fact that weaver is an ion channel suggests that this class of membrane receptor may be crucial during entry into a differentiation program, a program that includes migration, positioning and synaptogenesis.

Cell Surface Adhesion Systems in Migration

Neuroanatomical studies by Rakic (1971, 1972, 1978), including three-dimensional reconstruction of electron microscopic views, show that postmitotic neurons are closely apposed to the radial glial fibers. Thus, the radial glial fibe-system (Misson et al. 1991 a, b; Hunter and Hatten 1995) provides a primary substrate for migration in neocortex. In this model, young neurons are thought to use the glial fibers as a substrate for the journey out to the emerging layer of cortex, a distance of some 3,000 μm in higher vertebrates. A role for radial glia in directed migration along the radial plane of the neuroaxis was supported by studies in the developing hippocampus, where neurons followed the undulating pathways of the radial fibers to form the neural layers (Nowakowski and Rakic 1979). While McConnell and colleagues (O'Rourke et al. 1992) have recently shown that a minor population of cortical neurons migrate by other mechanism(s), more than 85% of the cells use the glial fiber system (Misson et al. 1991 a) as a substrate for directed migrations. Thus, the radial mode of migration is a cardinal feature of the developing cortex.

To examine the dynamics of neuronal migration of neurons along glial fibers, we have used video-enhanced differential contrast microscopy to obtain a detailed view of the morphology and behavior of migrating cerebellar neurons (Edmonson and Hatten 1987). The cytological features of living, migrating granule neurons in vitro were remarkably similar to those described by Rakic for cells that had been assumed to be migrating in vivo (Rakic 1971). Migrating neurons expressed a highly extended bipolar shape along the glial fiber, forming a close apposition with the glial process along the length of the neuronal cell soma and extending a leading process in the direction of migration (Hatten 1990, 1993; Gregory et al. 1988; Fishman and Hatten 1994; Rivas and Hatten 1994).

Astrotactin Provides a Ligand for Migration Along Glial Fibers

The use of purified cell populations combined with video microscopy thus provided both a detailed view of the movement of the granule cell along glial fibers and an assay system for the cell surface receptor systems that support these movements. Antibody perturbation studies on granule neuron migration in vitro demonstrated that the neural glycoprotein astrotactin provides a neural receptor system for migration along glial fibers (Edmondson et al. 1988; Fishell and Hatten 1991). Molecular cloning of the major component of the astrotactin activity indicates that the gene encodes a protein that contains EGF repeats and fibronectin type III domains (Zheng et al. 1996). By Northern analysis, cDNAs for astrotactin encode a brain-specific transcript that is developmentally regulated, showing high levels of expression in developing brain and low levels in adult brain.

To extend our work on developing murine brain to the problem of human cortical malformation, we used a 4Kb mouse cDNA probe to map astrotactin on both human and mouse metaphase chromosomes (Fink et al. 1997). A sequential high resolution G-band-to-fluorescence in situ hybridization (G to FISH) protocol localized the gene to 1q25.2 in humans and 1H2 in mice. The result with human chromosomal mapping indicates that astrotactin is distinct from genes mutated in human lissencephaly, mapped on 17p13.3 and Xq22, as well as cortical nodular heterotopia linked to Xq28. Thus astrotactin is a candidate gene for as yet unmapped neuronal migration disorders in man.

The location of astrotactin on the mouse chromosome confirms that astrotactin is not involved in either the reeler or weaver mouse mutations. The latter map to chromosomes 5 and 16, respectively. Work is in progress to map the murine astrotactin gene precisely (Millonig and Hatten, unpublished) and to examine animals with targeted mutations in the astrotactin gene (Dietz and Hatten, unpublished).

The Role of Migration in Neuronal Specification

Whereas attention has previously focused on the mechanism of glial guidance of neuronal positioning in developing brain, the central issues are shifting to how migration fits in an overall pattern of neuronal development. Is radial migration a means of providing a four-dimensional matrix in time and space of epigenetic regulatory elements? Purification and transplantation of progenitor cells from the developing cerebellar cortex into the embryonic cerebral cortex offer an approach to this question. Work is underway to examine the pathways of movement of EGL cells, harvested at different stages of development when different sets of genes are expressed, and phenotypic choices of transplanted EGL cells. Preliminary results indicate that granule cells retain some aspects of their developmental program, including the expression of a bipolar morphology. However, transplantation alters the specific patterns of gene expression and the survival of the cells within a heterotopic brain region (Li, Gallagher and Hatten, unpublished observations). This suggests that "subroutines" of genes, such as those needed for synaptogenesis with a particular target cell, or others, may exist within the overall program of gene expression. The influence of local regulating elements, such as the environment of the cortex, remains to be demonstrated.

RU49, A Zinc Finger Transcription Factor, is a Marker for Granule Cells

The role of cell type-specific transcription factors in lineage specification during invertebrate neural development (Jan and Jan 1993) is well established. Critical decisions concerning cell fate involve transcription factors that are expressed in the earliest precursors of a particular cell type. Often, these transcription factors continue to be expressed throughout the differentiation program for that cell type, providing definitive markers for a specific cell lineage in both the developing and adult organism. Recently, the Heintz laboratory has discovered a zinc finger-containing transcription factor that is expressed by granule cells, RU49 (Yang et al. 1996). Expression is restricted to granule cells within the cerebellum, allowing use of RU49 as a definitive marker for granule cells. Moreover, RU49 mRNA is detected in the site of origin of the EGL precursors and at all stages of granule cell development, including adulthood. The availability of this marker has made possible studies on mechanisms of granule cell precursor specification (Alder et al. 1996).

Conclusion

Understanding the molecular basis of cell migrations in developing brain is one of the key steps toward elucidation of the causes of brain malformations. Determining the molecular basis of cell movements and the role of the translocation of young neurons through a series of different local environments during the period of phenotypic specification, laminar positioning and synaptogenesis presents broad challenges for future research. The discovery of astrotactin and its role in migration, as well as of other genes that mark the changing patterns of gene expression during neuronal development, provides an initial step toward identifying genes that regulate the program of cortical formation.

Acknowledgments

We thank Christine Gallagher for help with the preparation of the manuscript. Supported by NIH grant NS-15429, NIH grant NS-34469 and The March of Dimes Birth Defect Foundation grant 1-FY95-0782 (MEH).

References

Alder J, Cho K, Hatten ME (1996) Embryonic precursor cells from the rhombic lip are specified to a cerebellar granule neuron identity. Neuron 17:389–399

Altman J, Bayer S (1985) Embryonic development of the rat cerebellum. I. Delineation of the cerebellar primordium and early cell movement. J Comp Neurol 231:1–26

Boulder Committee (1970) Embryonic vertebrate central nervous system: revised terminology. Anat Rec 166:257–262

Clapham DE (1994) Direct G protein activation of ion channels. Annu Rev Neurosci 17:441–464

Edmondson JC, Hatten ME (1987) Glial-guided granule neuron migration in vitro: a high-resolution time-lapse video microscopic study. J Neurosci 7:1928–1934

Doupnik CA, Davidson N, Lester HA (1995) The inward rectifier potassium channel family. Curr Opin Neurobiol 5:268–277

Edmondson JC, Liem RKH, Kuster JC, Hatten ME (1988) Astrotactin, a novel cell surface antigen that mediates neuron-glia interactions in cerebellar microcultures. J Cell Biol 106:505–517

Fink JM, Hirsch BA, Zheng C, Dietz G, Hatten ME, Ross ME (1997) Astrotactin (ASTN), a gene for glial-guided neuronal migration, maps to human chromosome 1q25.2. Genomics, 40:202–205

Fishell G, Hatten ME (1991) Astrotactin provides a receptor system for glia-guided neuronal migration. Development 113:755–765

Fishman R, Hatten ME (1993) Multiple receptor systems promote CNS neuronal migration. J Neurosci 13:3485–3495

Gao W-Q, Liu X-L, Hatten ME (1992) The weaver gene encodes a nonautonomous signal for CNS neuronal differentiation. Cell 68:841–854

Goldowitz D (1989) The weaver granuloprival phenotype is due to intrinsic action of the mutant locus in granule cells: evidence from homozygous weaver chimeras. Neuron 2:1565–1575

Goldowitz D, Mullen RJ (1982) Granule cell as a site of gene action in the *weaver* cerebellum: evidence from heterozygous mutant chimeras. J Neurosci 2:1474–1485

Gregory WA, Edmondson JC, Hatten ME, Mason CA (1988) Cytology and neuron-glia apposition of migrating cerebellar granule cells in vitro. J Neurosci 8:1728–1738

Hatten ME (1993) The role of migration in central nervous system neuronal development, Curent Opinions in Neurobiology, 3:38–44
Hatten ME (1990) Riding the glial monorail: a common mechanism for glial-guided migration in different regions of the developing brain. Trends Neurosci 13:179–184
Hatten ME (1992) The role of migration in central nervous system neuronal development. Curr Opin Neurobiol 3:38–44
Hatten ME, Heintz N (1995) Mechanisms of neural patterning and specification in developing cerebellum. Annu Rev Neurosci 18:385–408
Hatten ME, Liem RKH, Mason CA (1986) Weaver mouse cerebellar granule neurons fail to migrate on wild type astroglial processes in vitro. J Neurosci 6:2676–2683
Heginbotham L, Lu Z, Abramson T, MacKinnon R (1994) Mutations in the K^+ channel signature sequence. Biophys J 66:1061–1067
Heintz N, Norman DJ, Gao W-Q, Hatten ME (1993) Neurogenetic approaches to mannalian brain development. In: Davies KE, Tilghman SM (eds) Genome analysis Volume 6. Genome maps and neurological disorders. Cold Spring Harbor Press, Cold Spring Harbor, NY, pp 19–44
His JW, Ruffett TL (1889) Die Neuroblasten und deren Entstehung im embryonalen Mark. Abhand Math Phys Kl Konigl Sachs Ges Wiss 26:313–372
Hunter K, Hatten ME (1995) Radial glial cell transformation to astrocytes is bidirectional: Regulation by a diffusible factor in embryonic forebrain. Proc Natl Acad Sci USA 92:2061–2065
Jan YN, Jan LY (1990) Genes required for specifying cell fates in Drosophila embryonic sensory nervous system. Trends Neurosc 13:493–498
Kofuji P, Davidson N, Lester HA (1995) Evidence that neuronal G-protein-gated inwardly rectifying K^+ channels are activated by Gbg subunits and function as heteromultimers. Proc Natl Acad Sci USA 92:6542–6546
Kofuji P, Hofer M, Millen KJ, Millonig JH, Davidson N, Lester HA, Hatten ME (1996) Functional analysis of the weaver mutant GIRK2 potassium channel and rescue of weaver granule cells. Neuron 16:941–952
Krapivinsky G, Krapivinsky L, Velimirovic B, Wickman K, Navarro B, Clapham DE (1995) The cardiac inward rectifier K^+ channel subunit, CIR, does not comprise the ATP-sensitive K^+ channel, I_{KATP}. J Biol Chem 270:28777–28779
Kuhar SG; Feng L, Vidan S, Ross EM, Hatten ME, Heintz N (1993) Changing patterns of gene expression define four stages of cerebellar granule neuron differentiation. Development 117:97–104
Kurachi Y (1995) G protein regulation of cardiac muscarinic potassium channel. Am J Physiol 38:C821–C830
Miale I, Sidman RL (1961) An autoradiographic analysis of histogenesis in the mouse cerebellum. Expl Neurol 4:277–296
Miao GG, Smeyne RJ, D'Arcangelo G.n Copeland NG, Jenkins NA, Morgan JI, Curran T (1994) Isolation of an allele of reeler by insertional mutagenesis. Proc Nat Acad Sci USA 91:11050–11054
Misson J-P, Takahashi T, Caviness VS (1991 a) Early ontogeny of radial glial cells in the murine cerebral cortex. Glia 4:138–148
Misson J-P, Austin CP, Takahashi T, Cepko CL, Caviness VS Jr (1991 b) The alignment of migrating neural cells in relation to the murine neopallial radial glial fiber system. Cereb Cortex 1:221–229
North RA (1989) Drug receptors and the inhibition of nerve cells. Brit J Pharmacol 98:13–28
Nowakowski R, Rakic P (1979) The mode of migration of neurons to the hippocampus: a Golgi and electron microscopic analysis in fetal rhesus monkeys. J Neurocytol 8:697–718
O'Rourke NA, Dailey ME, Smith SJ, McConnell SK (1992) Diverse migratory pathways in the developing cerebral cortex. Science 258:299–302
Patil N, Cox DR, Bhat D, Faham M, Myers RM, Peterson AS (1995) A potassium channel mutation in weaver mice implicates membrane excitability in granule cell differentiation. Nature Genet 11:126–129
Pfaffinger PJ, Martin JM, Hunter DD, Nathanson NM, Hille B (1985) GTP-binding proteins couple cardiac muscarinic receptors to a K channel. Nature 317:536–538
Rakic P (1971) Neuron-glia relationship during granule cell migration in developing cerebellar cortex. A golgi and electron microscopic study in Macacus rhesus. J Comp Neurol 141:183–312

Rakic P (1972) Mode of cell migration to the superficial layers of foetal monkey neocortex. J Comp Neurol 145:61–84
Rakic P, Sidman RL (1973) Weaver mutant mouse cerebellum: defective neuronal migration secondary to abnormality of Bergmann glia. Proc Natl Acad Sci USA 70:240–244
Rakic P (1978) Neuronal migration and contact guidance in the primate telencephalon Postgrad Med J 54:25–40
Rakic P, Caviness VS (1995) Cortical development: view from neurological mutants two decades later. Neuron 14:1101–1104
Ramon y Cajal S (1889) Sobre las fibras nerviosas de la capa granulosa del cerebelo. Rev Trim de Histol Norm y Pathol 3 & 4:
Ramon y Cajal S (1911) Histologie du systeme nerveux de l'homme et des vertebres. Paris: Maloine (reprinted by Consejo Superior de Investigaciones Cientificas, Madrid, 1955).
Rezai Z, Yoon H (1972) Abnormal rate of granule cell migration in the cerebellum of "weaver" mutant mice. Dev Biol 29:17–26
Rivas R, Hatten ME (1994) Motility and cytoskeletal organization of migrating cerebellar granule neurons. J Neurosci 15:981–989
Sauer FC (1935) The cellular structure of the neural tube. J Comp Neurol 63:12–23
Schaper A (1897) The earliest differentiation in the central nervous system of vertebrates. Science 5:430–431
Sidman RL, Rakic P (1973) Neuronal migration with special reference to developing human brain: A review. Brain Res 62:1–35
Sotelo C, Changeux JP (1974) Bergmann fibers and granular cell migration in the cerebellum of homozygous weaver mutant mouse. Brain Res 77:484–491
Yang J, Jan YN, Jan LY (1995) Determination of the subunit stoichiometry of an inwardly rectifying potassium channel. Neuron 15:1441–1447
Yang X, Zhong R, Heintz N (1996) Granule cell specification in the developing mouse brain as defined by expression of the zinc finger transcription factor RU49. Development 122:555–566
Zheng C, Heintz N, Hatten ME (1996) CNS gene encoding astrotactin, a gene which supports neuronal migration along glial fibers. Science 272:417–419

Finding a Role for Cajal-Retzius Cells in Cerebral Cortex Development

C. Sotelo and E. Soriano

Introduction

Despite the important advances in unravelling the developmental history of the mammalian neocortex (Rakic 1977; O'Leary 1989), many unanswered questions still remain. Among these questions, one has been puzzling neuroscientists for over a century: what is the function of a distinct population of neurons known as the Cajal-Retzius cells (CR-cells)? These neurons, the first cortical cells to proliferate and differentiate, were reported by Cajal (1890) in the cortex of neonatal rodents and by Retzius (1891) in the developing human neocortex. CR-cells, together with subplate neurons (Allendoerfer and Shatz 1994), form the primordial plexiform layer or preplate. The subsequent appearance of the cortical plate, within the preplate, subdivides the latter into an outer and inner zone, which contain, respectively, mainly the CR-cells and the subplate neurons (Marin-Padilla 1971). Both categories of neurons have a transient life, and the majority undergo a process of cell death at the end of the neocortical ontogeny. Investigations carried out in the 1980s succeeded in demonstrating that subplate neurons have the important role of orienting extracortical afferent projections, particularly thalamocortical fibers, toward their still proliferating cortical target areas (Shatz et al. 1988; Ghosh et al. 1990; Allendoerfer and Shatz 1994). Subplate neurons are, therefore, essential for the formation and reshaping of thalamocortical afferents.

CR-cells remain in the marginal zone (future layer I) directly under the pial surface. They are composed of a large fusiform perikaryon, from which emerge horizontally-oriented dendrites and an extensive axonic arbor. The soma and dendrites contain all the usual organelles of mature neurons and receive synaptic inputs from en passant axons of unknown origin (Derer and Derer 1992). The mature appearance of the cytology of the CR-cells and their cytochemical differentiation (glutamate-like and calretinin-like immunoreactivity; Del Rio et al. 1995) testify that they are mature neurons. Owing to their transient life, CR-cells may be considered to be somewhat similar to some pioneer neurons in invertebrates and lower vertebrates that, after playing specific and essential roles during the formation of the nervous system, degenerate.

Newly generated neurons need to move along specific pathways to reach their final destination. This movement (neuronal cell migration) is a distinct cellular process that is essential for the establishment of the normal organization of the

A. M. Galaburda / Y. Christen (Eds.)
Normal and Abnormal Development of Cortex
© Springer-Verlag Berlin Heidelberg New York 1997

central nervous system. In particular, directional migration and precise spatiotemporal patterns of positioning are crucial for the correct formation of laminated brain regions such as the cerebellum and the neocortex (Hatten 1993; Rakic et al. 1994). Since the discovery by Rakic (1971) that radial glial fibers are one of the most common pathways of neuronal migration (glial-guided migration), special attention has been given to this glial population. Radial glia are among the first cells to differentiate, and they rapidly attain their specific phenotype. The latter is not only characterized by their bipolar shapes, forming spatially organized scaffolds, but also by the expression of specific markers, among them nestin and the antigens recognized by the RC1 and RC2 monoclonal antibodies (Hockfield and McKay 1985; Misson et al. 1988; Edwards et al. 1990). Radial glial cells are transient cells that are transformed into adult, mature astrocytes after neuronal migration is completed, when they lose the expression of their phenotypic markers. However, the cellular and molecular mechanisms that regulate radial glial identity and function are only partially known. Using cerebellar co-cultures of highly purified glia and neurons, Hatten (1985) has shown that the flattened forms of astroglia were transformed into elongated radial-like glial in the presence of neurons. These *in vitro* observations, indicating that neurons influence the differentiation of astroglia, have recently been extended (Hunter and Hatten 1995) by showing that the expression of radial glia phenotype required extrinsic soluble signals present in embryonic forebrain.

One of the putative functions of CR-cells, hypothesized years ago by Marin-Padilla (1984), is that they regulate the inside-out migration of cortical neurons. This suggestion was based upon the strategic location of CR-cells in layer I, where all migrating neurons are attracted, and all of them may contact CR-cells (Del Rio et al. 1995). The recent cloning of the *reeler* gene (D'Arcangelo et al. 1995) has provided further support for this hypothesis. *Reeler* is an autosomal recessive mutation that affects neuronal migration (Rakic and Caviness 1995). The pattern of the cerebral cortex in this mutant mouse is inverted in relationship to the normal cortex; therefore, the inside to outside order of neurogenesis that characterizes the normal organization of the neocortex does not occur in the *reeler* mutant. The wild-type gene at the *reeler* locus encodes Reelin, an extracellular matrix-like protein with eight EGF motifs and an amino-terminal region exhibiting some homology with F-spondin (D'Arcangelo et al. 1995; Hirotsune et al. 1995). From the *reeler* mouse, Ogawa et al. (1995) have isolated an alloantibody by immunizing homozygotes with homogenates taken from the neocortex of normal mouse embryos, and obtained the monoclonal antibody CR-50, which recognizes the Reelin protein (D'Arcangelo et al. 1997). During early neocortical development, Reelin mRNA and protein are only expressed in CR-cells. Furthermore, using dissociation-reaggregation cultures of neocortex and mab CR-50 to block the function of Reelin, Ogawa et al. (1995) were able to transform the normal histotypic organization of reaggregation cultures into the *reeler* mutant pattern of organization.

It is, therefore, most probable that CR-cells are involved in the migration of cortical neurons. However, the cellular mechanisms by which CR-cells are implicated in this migratory process remain unknown. Do CR-cells regulate astroglial morphology and the acquisition and maintenance of radial glia phenotype? Do CR-cells act through direct cell-to-cell contacts or mainly via secreted soluble factors? Are these pioneer neurons involved in the directional attraction of the migrating cells? The knowledge of one of us (ES) about the development of the cortex, particulary of CR-cells in mouse embryos, and the experience of the other (CS) with problems related to cerebellar transplantation prompted us to collaborate. We have carried out heterochronic and heterotopic transplantations, using the cerebellum as the biological system to test presumptive roles of CR-cells in the induction and maintenance of radial glia phenotype. In addition, we have performed *in vitro* experiments, with neocortical and cerebellar explants, to determine whether CR-cells are able to secrete diffusible molecules influencing cerebellar astroglial cells, and whether these neurons can attract migrating granule cells. The results of all these experiments have been recently published (Soriano et al., 1997).

Validation of the Adult Mouse Cerebellum as a System to Study Neuronal Regulation of Radial Glia Phenotype

Embryonic cerebellar neurons can be grafted into adult deficient cerebellum and become functionally integrated into the synaptic circuitry of the host cerebellar cortex (Sotelo and Alvarado-Mallart 1987 a, b). This integration occurs because grafted Purkinje cells are able to leave the grafts and migrate radially into the host molecular layer (Sotelo et al. 1990). Studies of transplants from a transgenic mouse line (Krox-20/lacZ14), in which Bergmann fibers could be identified by transgene expression, have shown that 1 or 2 months after grafting there was β-galactosidase activity in Bergmann fibers of the graft remnant. However, no β-galactosidase activity could be detected in the molecular layer of the host, even if the layer contained numerous grafted Purkinje cells (Sotelo et al. 1994). These results demonstrate the lack of comigration of embryonic Purkinje cells and embryonic Bergmann fibers, and corroborate the previously advanced hypothesis (Sotelo et al. 1990) that grafted Purkinje cells reach their location in the host molecular layer by radial migration along adult Bergmann fibers.

Although little is known about the molecular identity of the signals involved in the glial-guided neuronal migration (Rakic et al. 1994), these signals must be developmentally regulated and transiently expressed. During ontogeny, the neuronal and glial expression of the necessary signals is coordinated among the participant cells – premigratory neurons and their corresponding glial axes – which have closely related ages. The problem is entirely different in our transplantation experiments, where embryonic neurons interact with adult astrocytes. Under these conditions, glial-guided neuronal migration is possible only if the grafted neurons regulate the astroglia phenotype and induce the transient re-expression of radial glia. This idea, a "rejuvenation" of adult host Bergmann fibers by embryonic Purkinje cells, has been tested by checking for the re-expression of

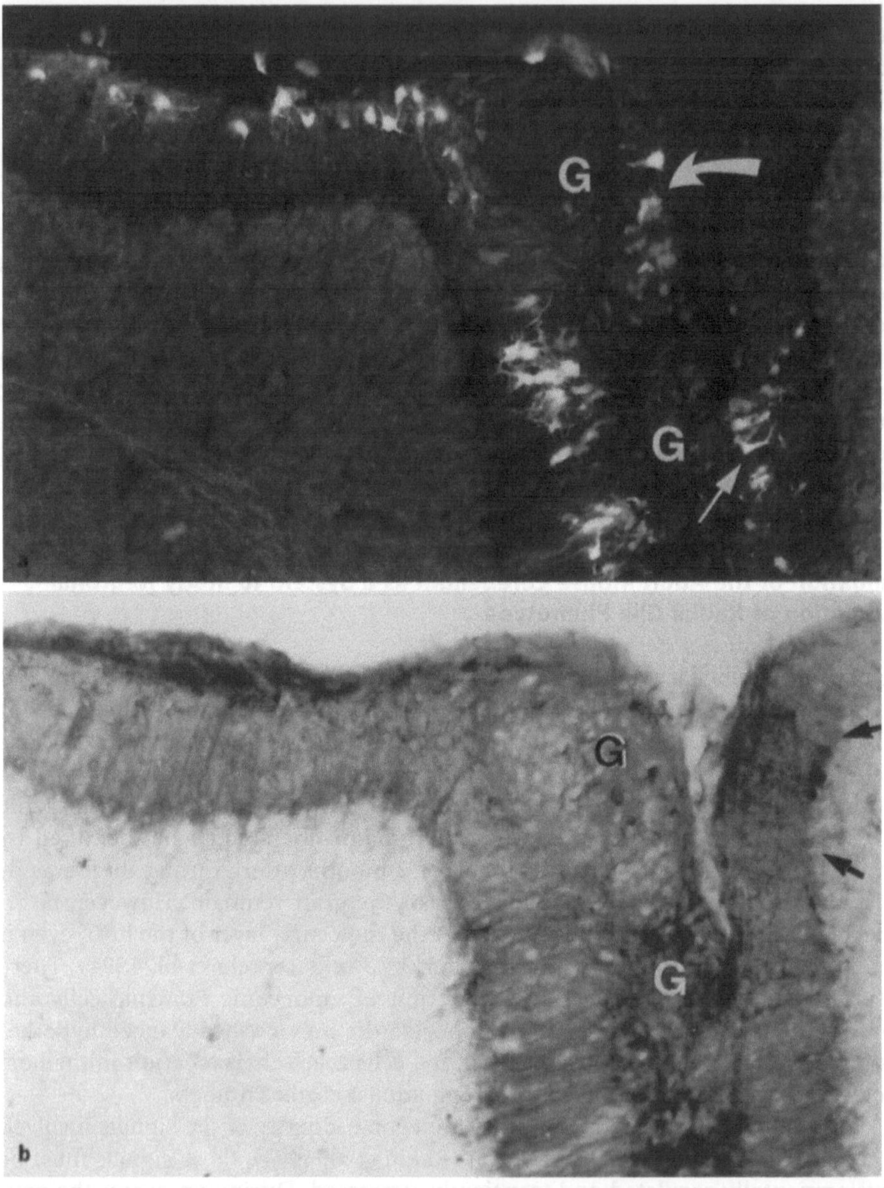

Fig. 1. Parasagittal section of the cerebellum of an adult pcd/pcd mutant mouse seven days after transplantation of E12 wild-type mouse cerebellar primordium. The section has been double-immunostained with a polyclonal antiserum against calbindin (CaBP; fluorescent) and the mab Rat-401 (diamino-benzidine). **A**, the molecular layer of the host contains CaBP+ Purkinje cells in their radial migration. These grafted neurons are numerous in the folium at the left of the micrograph, whereas they are scarce in the folium at the right (arrow). Note the occurrence of the graft remnant (G) between both folia, and the presence of CaBP+ cells in the graft remnant (curved arrow). **B**, the two folia exhibit intense Rat-401 immunoreactivity in their molecular layers. This immunostaining belongs to host Bergmann fibers, which are numerous in both folia even in some regions without grafted Purkinje cells (arrows). Rat-401-positive elements occur also in the graft remnant (G). x 160

nestin (an intermediate filament normally present only during the active phase of neuronal migration and, therefore, one of the radial glia markers) in adult Bergmann fibers using immunostaining with the mab Rat-401 (Hockfield and McKay 1985). Adult host Bergmann fibers re-expressed nestin during the entire time window of radial migration of grafted Purkinje cells; that is to say, from 6 to 12 days after grafting (DAG). There was often an excellent correlation between the density of migrating Purkinje cells and the intensity of Rat-401 staining in the apposed Bergmann fibers. Nevertheless, in some instances the occurrence of a few migrating Purkinje cells (Fig. 1 A) sufficed to induce intense antigen expression in patches of host Bergmann fibers (Fig. 1 B). Thus, the regulation of the radial glia phenotype did not occur as a result of a direct cell-to-cell interaction but, most probably, by way of diffusible factors secreted by the embryonic neurons (Sotelo et al. 1994).

Correct migration of grafted embryonic Purkinje cells in adult host cerebellum occurs because mature astrocytes are transiently transformed into juvenile radial glia. This transformation seems to take place in response to diffusible signals produced by the grafted embryonic neurons. These results provide the first *in vivo* evidence in favor of a neuronal influence in the differentiation of astroglial cells. Moreover, they validate the adult cerebellum and heterotopic and heterochronic transplantations as an *in vivo* system in which to test whether different populations of embryonic neurons have or do not have the capability to transform adult Bergmann fibers into cells with a radial glia phenotype. This system was used in our investigation of the role of CR-cells.

Cajal-Retzius cells Induce the Transformation of Adult Bergmann Fibers into Astroglial Cells with Radial Glia Phenotype: *in vivo* Heterochronic and Heterotopic Transplantations

In the first series of experiments, solid transplants taken form the cortical anlage of mouse embryos 15 to 16 days postconception (E15, E16) were transplanted into adult mouse cerebellum. The implanted mice were examined 7 to 10 DAG. The implants did not invade the cerebellar parenchyma, and evolved into large masses of cortical tissue, which remained in the cerebellar fissures between the folia (Fig. 2 A). In sections immunostained with the mab RC2 (another marker of radial glia; Misson et al. 1988), zones of molecular layer containing RC2-positive palisades of Bergmann fibers occurred only in areas apposed to the neocortical grafts (Fig. 2 B). Parallel sections, immunostained with antibodies against calretinin (a calcium-binding protein marker of CR-cells; Del Rio et al. 1995) revealed clusters of surviving CR-cells within the grafts (Fig. 1 A). In serial reconstructions, these clusters of CR-cells happened to be either apposed or within short distances of the RC2-positive Bergmann fibers. These results support but do not prove the presumed regulation of astroglia phenotype by CR-cells, because many other cell populations co-exist within the cortical grafts.

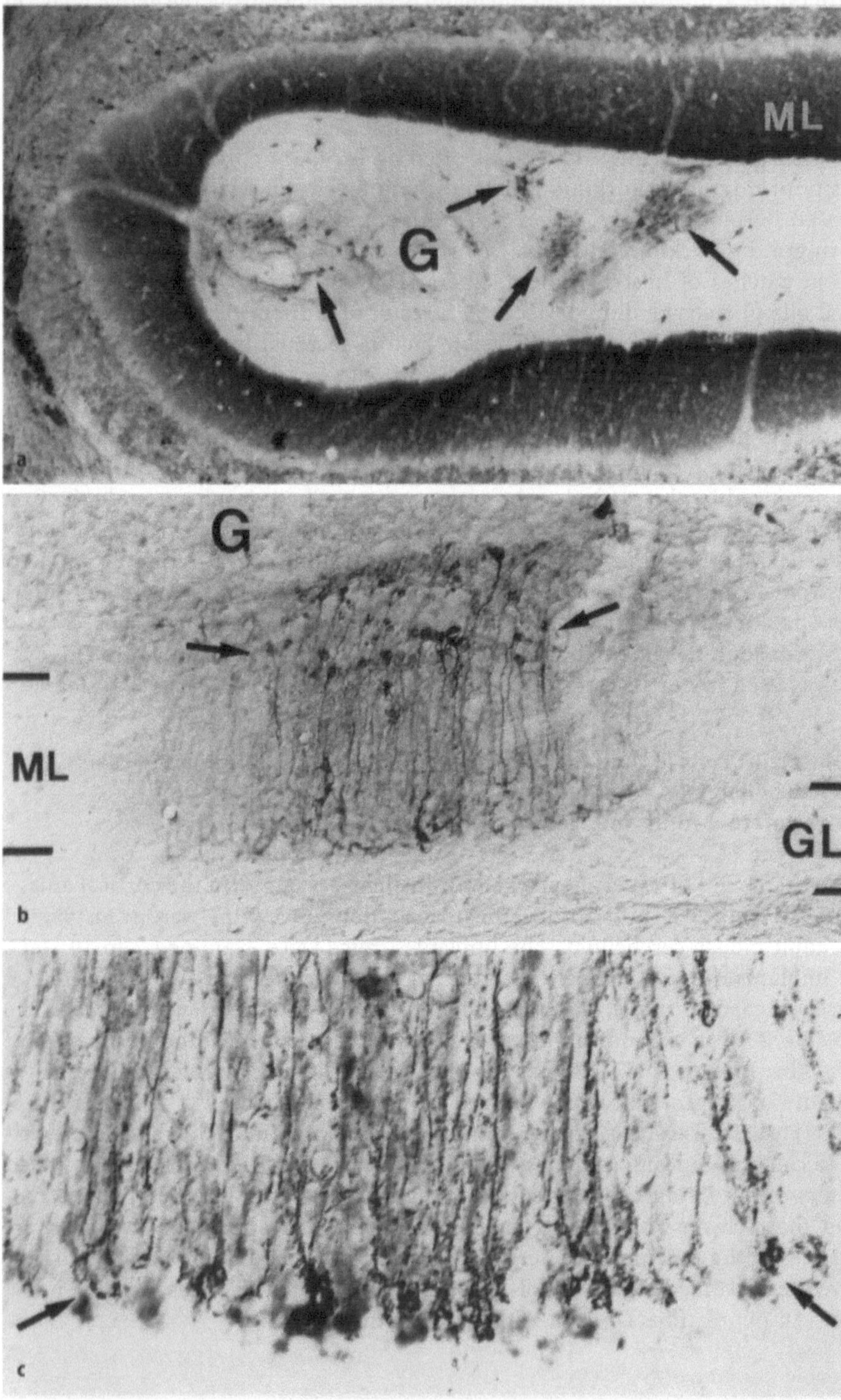
ML
G
a
G
ML
GL
b
c

◀ **Fig. 2.** Induction of RC2 immunoreactivity in Bergmann glia of host adult cerebellum grafted with either the whole E15 neocortical primordium (**B**) or with Cajal-Retzius cell-enriched grafts (**C**). **A**, calretinin immunostaining in a cerebellum grafted with the whole neocortical primordium, seven days after transplantation. This parasagittal section illustrates the survival of Cajal-Retzius cells within the graft. Note that the graft (G) does not invade the cerebellar parenchyma, and that Cajal-Retzius cells remain in small clusters (arrows), some of them close to the host molecular layer. The latter shows a strong calretinin immunoreactivity of their parallel fibers. **B**, small patch of RC2-positive Bergmann glia apposed to the neocortical graft (G). Note that they invade the neocortical graft (arrows). ML, molecular layer; GL, granule cell layer. **C**, induction of RC2 immunoreactivity seven days after transplantation of Cajal-Retzius cell-enriched slices. This microphotograph illustrates a large palisade of intense RC2+ Bergmann fibers (arrows) spanning the whole molecular layer. Note that this patch is larger than that induced by the whole neocortical primordium (**C**, **B**). **A**, x 90; **B**, x 350; **C**, x 600

To determine whether CR-cells are responsible for the observed Bergmann fibers' changes, two other series of transplantation experiments were carried out. First, we developed a simple method to enrich CR-cells by obtaining a tangential vibratome section from the flattened cortical subpial surface. This section, 50–70 μm thick, contained the whole layer I and, therefore, was enriched in CR-cells. The grafting of CR-cell-enriched slices into adult cerebellum gave rise to smaller implants, which also remained between cerebellar folia, with CR-cells (checked with calretinin-immunostaining) lying over the surface of the cerebellum. This kind of grafting gave rise to a strong induction of RC2 in the underlying Bergmann fibers (Fig. 2 C). Not only was the intensity of the RC2 immunoreaction stronger than in cases with whole cortical transplants, but the number of RC2-positive patches of Bergmann fibers was also increased (Fig. 2 C).

Finally, in control experiments, we used the remaining cortical anlage - after removing the first vibratome slice - as donor tissue. The latter, lacking CR-cells, did not induce RC2 immunoreactivity in any host Bergmann fibers. These three series of heterochronic and heterotopic transplantations allowed us to conclude that CR-cells, but not other categories of cells present in E15–E16 cortical anlage, can induce the transformation of mature Bergmann fibers into juvenile radial glia. Thus, CR-cells have the functional role of influencing the differentiation of radial glial cells.

Cellular Mechanisms by which Cajal-Retzius Cells are Involved in Neuronal Cell Migration: *in vitro* Studies

The *in vivo* experiments described above strongly suggest that CR-cells exert a regulatory influence on adult cerebellar Bergmann glia. However, these experiments do not help to unravel the cellular mechanisms implicated in the neurono-glial regulation and, therefore, do not allow us to answer most of the questions raised in our Introduction. With this aim in mind, we have performed three series of *in vitro* experiments planned to investigate the following issues.

1) The Morphogenetic Influence of Cajal-Retzius Cells on Developing Glial Cells

Previous studies on the spatiotemporal pattern of RC2 expression in postnatal cerebellum have shown that RC2 immunostaining disappears from cerebellar radial glia at the end of the first postnatal week (~ P8: Edwards et al. 1990; Misson et al. 1988). To determine the influence of CR-cells on developing glial cells we used a confrontational *in vitro* assay. This assay consisted of cerebellar slice cultures taken from P6–P8 Krox-20/lacZ14 mice positioned in the same plane between two opposed wild-type neocortical slices(R15–E16). The neocortical slice on one side was taken from layer I (CR-cell-enriched slices), and on the other, control side, it was taken from the ventricular zone or the cortex-minus-CR-cell preparations. The two neocortical slices were apposed to the cerebellar external granular layer. After 7–14 DIV, the co-cultures were stained for βgal activity to identify cerebellar Bergmann fibers. The CR-cell-enriched slice exerted a strong growth-promoting, morphogenetic influence on cerebellar Bergmann fibers, since there was a massive ingrowth of Xgal-positive fibers that penetrated the CR-cell slice for distances as long as 600 μm. In contrast, the glial cell ingrowth into the neocortical explants devoid of CR-cells was much more moderate, and the invasion distances were never greater than 200 μm. Thus, CR-cells also have an effect on developing glial cells and, through this growth-promoting influence, are implicated in the organization of the radial glia scaffolds necessary for neuronal cell migration.

2) The Inductive Action of Cajal-Retzius Cells on Cerebellar Glia is Mediated by Diffusible Factors

In two previous studies (Sotelo et al. 1994; Hunter and Hatten 1995) reporting a neuronal role in the induction and maintenance of radial glia phenotype, the neurons were supposedly acting through secretion of soluble and diffusible signals. To determine whether CR-cells also exert their action by a similar mechanism, we used an *in vitro* diffusion test assay. The wild-type cerebellar slices (P6–P8) were mounted onto Millicell CM membranes and were overlaid with a semipermeable membrane. CR-cell-enriched explants or control neocortical explants lacking CR-cells were positioned over the semipermeable membrane, just above the cerebellar explants. After 7 DIV, the cerebellar explants were examined for RC2 immunoreactivity. In the cerebellar explants overlaid with CR-cells, there was a dramatic induction of RC2 immunoreactivity, not only in the molecular layer where Bergmann fibers are located but throughout the cerebellar cortex and white matter (Fig. 3 A). The majority of these glial cells exhibited elongated bipolar shapes with two processes emerging from opposite ends (Fig. 3 B, C), resembling very immature radial glia in the developing brain (Rakic 1971; Edwards et al. 1990). Conversely, in the cerebellar explants overlaid with control neocortical explants devoid of CR-cells, not one RC2-positive cell occurred.

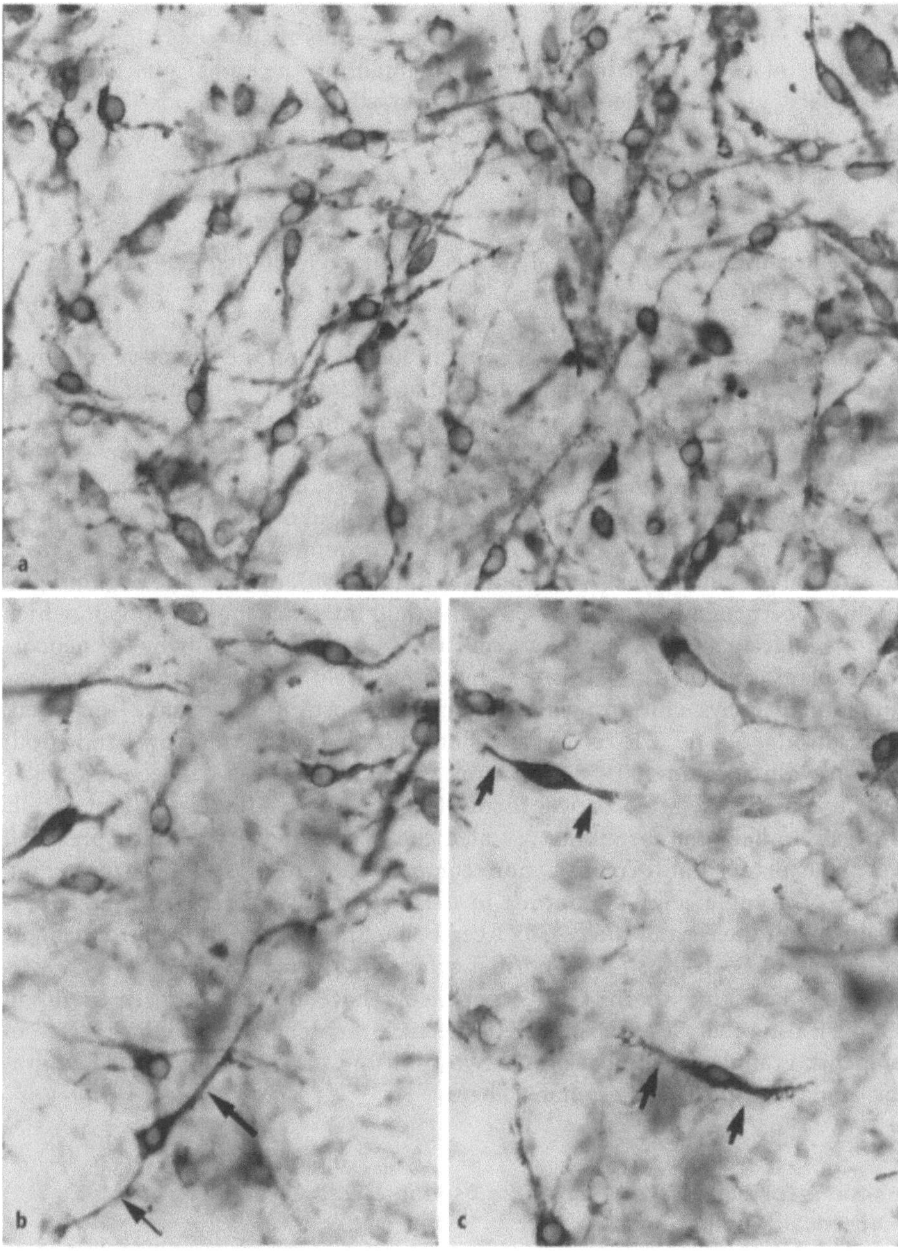

Fig. 3. Induction of RC2 immunoreactivity in postnatal cerebellar slices co-cultured with a Cajal-Retzius cell-enriched slice separated by a semipermeable membrane (diffusion co-culture assay). **A**, note the impressive induction of RC2 immunoreactivity in numerous cerebellar glial cells. These cells, occurring throughout the cerebellar slice, appear as bipolar, randomly oriented cells. **B**, some of the RC2-positive cells have asymmetric processes, one being thicker and longer than the other (arrows). **C**, this micrograph illustrates very immature cells with two thick and short processes emerging from opposite poles of the cell somata (arrows). **A, B, C**, x 185

The *in vitro* diffusible test assays provided results that strongly indicate that the action of CR-cells in inducing and maintaining the radial glia phenotype takes place via diffusible factors. Moreover, this action is dose-dependent, because the degree of the radial glia induction is determined by the concentration of these unidentified diffusible factors.

3) The Attractive Influence of Cajal-Retzius Cells upon Migrating Cerebellar Granule Cells

To test whether CR-cells may exert some influence upon the direction of movement of the neurons migrating along radial glial axes, we used the triple co-cultures confrontational assays reported above. In these instances, the cerebellar slices were not taken from Krox-20/lacZ14 transgenic mice but from another strain, the β2nZ3'1. In the cerebellum of these mice, βgal shows ectopic expression in granule, basket and stellate cells (Fig. 4 A; Jankovski et al. 1996); of these categories of neurons only granule cells, the most numerous, are generated from the external granular layer (Zhang and Goldman 1996), up until the beginning of the third postnatal week. Thus, if CR-cells could influence the direction in which neurons migrate, cerebellar granule cells would be attracted only by the neocortical explants enriched in CR-cells. This is precisely what happens in our confrontational assays cultures, where numerous postmitotic granule cells were found throughout the CR-cell explant but, particularly, in a narrow band about 200 μm wide, located 400–500 μm away from the interface between cerebellar and CR-cell-enriched neocortical tissues (Fig. 4 B). The visualization of CR-cells with calretinin immunostaining revealed the occurrence of CR-cells spread all over the explants, but forming a more compact band in those zones with a high density of granule cells. In contrast to these changes in cerebellar granule cell organization, the area of the β2nZ3'1 cerebellar slice apposed to control neocortex, devoid of CR-cells, showed normal cytoarchitecture.

These last kind of *in vitro* experiments have been extremely useful in demonstrating that CR-cells are able to perturb the normal migratory pattern of cerebellar granule cells, most probably by shifting the direction of their migration along RC2-positive Bergmann fibers.

Functions for Cajal-Retzius Cells Disclosed in Mixed Cerebellar and Neocortical Preparations

The CR-cells, but no other cortical cells, are responsible for the transformation of Bergmann glia into radial glia, both in adult and postnatal cerebellum. Thus, the radial glia-to-astrocyte pathway is a bidirectional process that is regulated by soluble factors produced by CR-cells. Moreover, these factors also exert a growth-promoting effect on postnatal Bergmann fibers, and induce a radial glia phenotype in most astrocytes in the postnatal cerebellum. Neuronal-to-radial glia

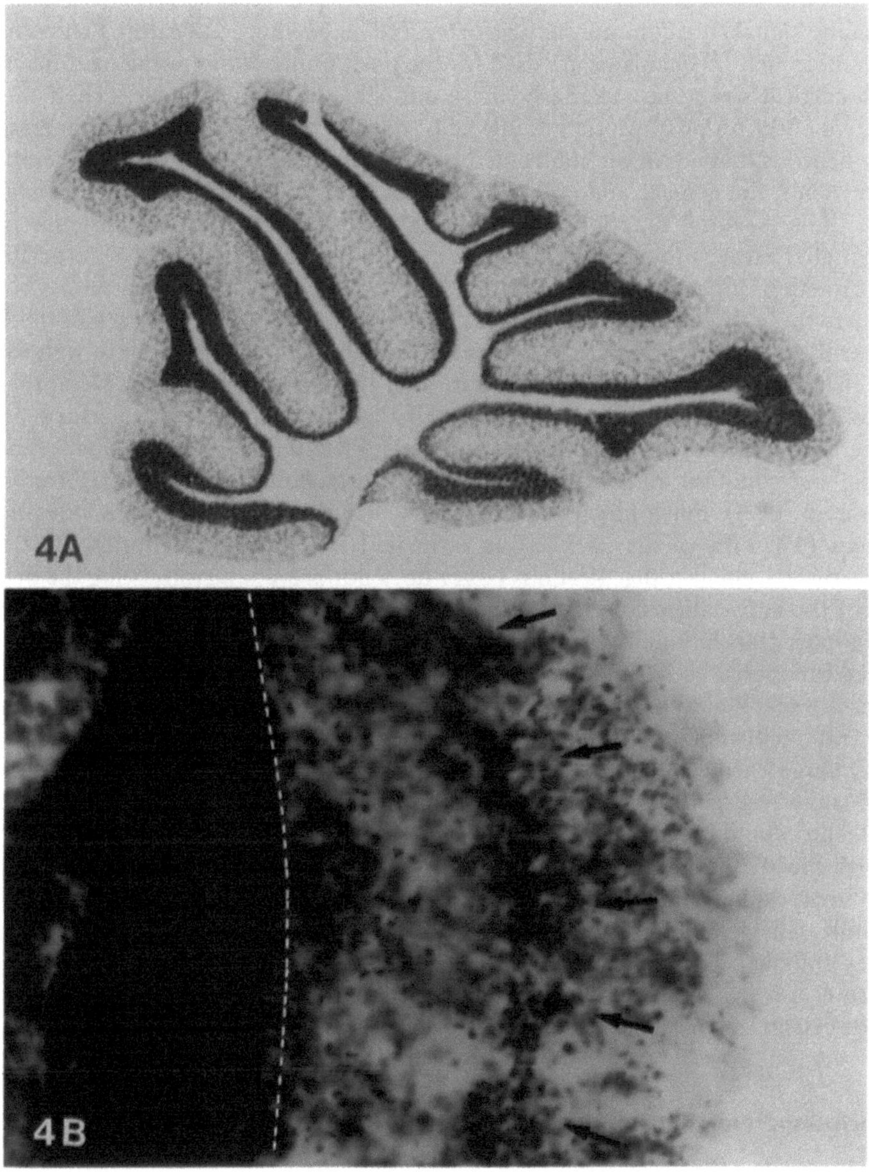

Fig. 4. Color micrographs illustrating the β2nZ3'1 transgenic mouse cerebellum, and the attractive effect of Cajal-Retzius cell-enriched slice on migrating granule cells. **A**, parasagittal section of a P21 cerebellum taken from a β2nZ3'1 transgenic mouse. βgal activity has been revealed with X-gal, and owing to the "nls/lacZ" construct, reaction product occurs in neuronal nuclei. Cerebellar cortical neurons expressing the reporter gene are only granule cells and molecular layer interneurons (stellate and basket cells). **B**, high magnification of a Cajal-Retzius cell-enriched slice that has been co-cultured (in a triple co-culture confrontational assay) with a β2nZ3'1 cerebellar slice. The broken line marks the border between cerebellum and Cajal-Retzius cell slice. Note that βgal-positive granule cells have massively invaded the Cajal-Retzius cell-enriched slice. The arrows point to a band of densely packed granule cells at a distance from the slices interface. **A**, x 25; **B**, x 150

inducing signals are common to neocortex and cerebellum, indicating that common molecular mechanisms are used for the positioning of migrating neurons in both cortical structures. Similarly, these diffusible factors also have a chemoattractive effect on postmitotic granule cells, despite being postmitotic premigratory neurons that are never confronted by CR-cells during their normal development. Thus, the generic and still unknown factors produced by CR-cells have at least three essential functions: 1) to regulate the formation and persistence of radial glial axes; 2) to promote the growth of radial glial fibers; and 3) to orient the direction of neuronal cell migration.

Are these three functions regulated by the same molecule, or is each of them under the control of a specific molecule? Although the final answer must wait for the characterization of the chemical nature of the diffusible factors, several theories can be advanced. One possibility is that a single molecule can perform all these functions. For sure, the main candidate is Reelin, owing to the phenotype of the *reeler* mutant mouse. Reelin has a very large molecular weight (D'Arcangelo et al. 1995), much larger than the molecular weight of radial glia-inducing signals (390 kDa versus 55 kDa) as estimated by Hunter and Hatten (1995). Nevertheless, the fact that Reelin is larger does not prevent it from being the origin of the active diffusible factor. Indeed, a processed form of Reelin - one of many post-translationally processed forms of the protein - could have the proposed functional activity. However, it looks more likely that several active molecules are secreted by CR-cells. In this hypothesis, Reelin, which has a low diffusion rate, will remain within a short distance of CR-cells. Reelin would act, as other large extracellular matrix proteins do for some growth factors (Lander 1993), as a reservoir of other diffusible molecules. The latter, remaining attached to Reelin, would create a distinct zone - corresponding to the neocortical layer I - with specific induction and chemoattractant properties. This molecular diversity might explain some features of the *reeler* phenotype: for instance, the fact that granule cells - although greatly reduced in number - migrate normally to the inner granule cell layer; the occurrence of RC2-positive radial glia; or the occurrence of a low percentage of Purkinje cells that have correctly migrated at the molecular/granular layer interface in the *reeler* cerebellum (Mariani et al. 1977).

Conclusions: Functions of Cajal-Retzius Cells in Neocorticogenesis

The results summarized in this chapter allow us to conclude that CR-cells - the first-generated cortical neurons - fulfill two specific and essential functions: they are implicated in the organization of the radial glial palisades, and they seem to attract migrating neurons. Thus, these two complementary functions guarantee the "inside-out" order of migration and corticogenesis. Nevertheless, the important roles played by CR-cells in neuronal cell migration are not incompatible with other presumptive functions. Thus, recent results by Del Rio et al. (1997) obtained in co-cultured hippocampus and entorhinal cortex strongly suggest that CR-cells, similarly to subplate neurons (the other early generated and transient neuronal population of the neo- and allocortices; Allendoerfer and Shatz 1994), also act as signposts to guide efferent axons to their postsynaptic targets.

References

Allendoerfer KL, Shatz CJ (1994) The subplate, a transient neocortical structure: its role in the development of connections between thalamus and cortex. Ann Rev Neurosci 17:185–218

Cajal SR (1890) Sobre la existencia de células nerviosas especiales en la primera capa de las circunvoluciones cerebrales. Gazeta Médica Catalana 13:737–739

D'Arcangelo G, Miao GG, Chen S-C, Soares HD, Morgan JI, Curran T (1995) A protein related to extracellular matrix proteins deleted in the mouse mutant reeler. Nature 374:719–723

D'Arcangelo G, Nakajima K, Miyata T, Ogawa M, Mikoshiba K, Curran T (1997) Reelin is a secreted glycoprotein recognized by the CR-50 monoclonal antibody. J Neurosci 17:23–31

Del Río JA, Martínez A, Fonseca M, Auladell C, Soriano E (1995) Glutamate-like immunoreactivity and fate of Cajal-Retzius cells in the murine cortex as identified with calretinin antibody. Cereb Cortex 5, 13–21

Del Río JA, Heimrich B, Borrell V, Förster E, Drakew A, Alcántara S, Nakajima K, Miyata T, Ogawa M, Mikoshiba K, Derer P, Frotscher M, Soriano E (1997) A role for Cajal-Retzius cells and reelin in the development of hippocampal connections. Nature 385:70–74

Derer P, Derer M (1992) Development and fate of Cajal-Retzius cells in vivo and in vitro. In: Sharma SC, Goffinet AM (eds) Development of the central nervous system in vertebrates. New-York, Plenum Press, pp 113–127

Edwards MA, Yamamoto M, Caviness VS (1990) Organization of radial glial and related cells in the developing murine CNS: An analysis based upon a new monoclonal antibody marker. Neuroscience 36:121–144

Ghosh A, Antonini A, McConnell SK, Shatz CJ (1990) Requirement for subplate neurons in the formation of thalamocortical connections. Nature 347:179–181

Hatten ME (1985) Neuronal regulation of astroglial morphology and proliferation in vitro. J Cell Biol 100:384–396

Hatten ME (1993) The role of migration in central nervous system neuronal development. Curr Opin Neurobiol 3:38–44

Hirotsune S, Takahara T, Sasaki N, Hirose K, Yoshiki A, Ohashi T, Kusakabe M, Murakami Y, Muramatsu M, Watanabe S, Nakao K, Katsuki M, Hayashizaki Y (1995) The reeler gene encodes a protein with an EGF-like motif expressed by pioneer neurons. Nat Genet 10:77–83

Hockfield S, McKay R (1985) Identification of major cell classes in the developing mammalian nervous system. J Neurosci 5:3310–3328

Hunter KE, Hatten ME (1995) Radial glial cell transformation to astrocytes is bidirectional: regulation by a diffusible factor in embryonic forebrain. Proc Natl Acad Sci USA 92:2061–2065

Jankovski A, Rossi F, Sotelo C (1996) Neuronal precursors in the postnatal mouse cerebellum are fully committed cells: evidence from heterochronic transplantations. Eur J Neurosci 8:2308–2319

Lander AD (1993) Proteoglycans in the nervous system. Curr Opin Neurobiol 3:716–723

Mariani M, Crepel F, Mikoshiba K, Changeux JP, Sotelo C (1977) Anatomical, physiological and biochemical studies of the cerebellum from *reeler* mutant mouse. Phil Trans Royal Soc Lond B 281:1–28

Marín-Padilla M (1971) Early prenatal ontogenesis of the cerebral cortex (neocortex) of the *Felix domestica*. A Golgi study. I. The primordial neocortical organization. Z Anat Entwicklungsgesch 134:117–145

Marín-Padilla M (1984) Neurons of layer I. A developmental analysis. In: Peters A, Jones EG (eds) Cerebral cortex. Vol I. Cellular components of the cerebral cortex. New York, Plenum Press, pp 447–478

Misson JP, Edwards ME, Yamamoto M, Caviness VS (1988) Identification of radial glial cells within the developing murine central nervous systems: studies based upon a new histochemical marker. Dev Brain Res 44:95–108

Ogawa M, Miyata T, Nakajima K, Yagyu K, Seike M, Ikenaka K, Yamamoto H, Mikoshiba K (1995) The reeler gene-associated antigen on Cajal-Retzius neurons is a crucial molecule for laminar organization of cortical neurons. Neuron 14:899–912

O'Leary DDM (1989) Do cortical areas emerge from a protocortex? Trends Neurosci 12:400–406

Rakic P (1971) Neuron-glia relationship during granule cell migration in developing cerebellar cortex. A Golgi and electron microscopic study in Macacus rhesus. J Comp Neurol 141:282–312

Rakic P (1977) Prenatal development of the visual system in rhesus monkey. Phil Trans Royal Soc Lond B 278:245–260

Rakic P, Caviness VS Jr (1995) Cortical development: view from neurological mutants two decades later. Neuron 14:1101–1104

Rakic P, Cameron RS, Komuro H (1994) Recognition, adhesion, transmembrane signaling and cell motility in guided neuronal migration. Curr Opin Neurobiol 4:63–69

Retzius G (1891) Über den Bau der Oberflächenschicht der Grosshirnrinde bei Menschen und bei Säugetieren. Verh Biol 3:90–103

Shatz CJ, Chun JJM, Luskin MB (1988) The role of the subplate in the development of the telencephalon. In: Jones EG, Peters A (eds) The cerebral cortex. Vol. II. The development of the cerebral cortex. New York, Plenum pp 35–58

Soriano E, Alvarado-Mallart RM, Dumesnil N, Del Rio JA, Sotelo C (1997) Cajal-Retzius cells regulate the radial glia phenotype in the adult and developing cerebellum and alter granule cell migration. Neuron 18:563–577.

Sotelo C, Alvarado-Mallart RM (1987 a) Reconstruction of the defective cerebellar circuit in adult Purkinje cell degeneration mutant mice by Purkinje cell replacement through transplantation of solid embryonic implants. Neuroscience 20:1–22

Sotelo C, Alvarado-Mallart RM (1987 b) Embryonic and adult neurons interact to allow Purkinje cell replacement in mutant cerebellum. Nature 327:421–423

Sotelo C, Alvarado-Mallart RM, Gardette R, Crepel F (1990) Fate of grafted embryonic Purkinje cells in the cerebellum of the adult "Purkinje cell degeneration" mutant mouse. I. Development of reciprocal graft-host interactions. J Comp Neurol 295:165–187

Sotelo C, Alvarado-Mallart RM, Frain M, Vernet M (1994) Molecular plasticity of adult Bergmann fibers is associated with radial migration of grafted Purkinje cells. J Neurosci 14:124–133

Zhang L, Goldman JE (1996) Generation of cerebellar interneurons from dividing progenitors in white matter. Neuron 16:47–54

Intra- and Extracellular Control of Neuronal Migration: Relevance to Cortical Malformations

P. Rakic

Abstract

Although neuronal cell migration as a major developmental phenomenon has been known since the end of the last century, it did not receive much attention until the advent of 3H-thymidine autoradiography and electron microscopy, which enabled determination of the rate and pathways of migrating neurons and examination of their ultrastructure and relationship to surrounding cells *in situ*. These studies led to concepts of inside out gradient of neurogenesis, neuron-glial cell differential adhesion, the radial unit hypothesis of cortical development and the protomap hypothesis of cortical parcellation.

More recently, methods of cellular biology and various *in vitro* approaches have provided an unprecedented opportunity to study the molecular mechanisms of neuronal migration. These studies revealed that directed, glia-guided neuronal migration requires the orchestration of multiple molecular events. These include selection of migratory pathways by cell recognition receptors, formation of adhesive interactions with cellular and extracellular substrates, and activation of specific ion channels that provide second messenger-mediated signals for the cytological changes involved in motility. Understanding the role of individual molecular components that mediate neuronal migration provides new insight into the pathogenesis of congenital cortical malformations.

Introduction

Since all cortical neurons originate near the surface of the embryonic cerebral ventricle, they all must move to their final positions in the cortex, which develop within the outer territories of the cerebral wall, below the pia. The migratory pathways of individual neurons become particularly long in the large primate telencephalon. Since the size of a migratory neuron in the mouse and human embryo is approximately the same, the negotiation of the tortuous, several thousand microns-long pathway in the gyrencephalic brain poses a daunting problem. Indeed, a massive migration of neurons in primates, including humans, occurs during midgestation and coincides with the rapid increase in the width of the cerebral wall and the buckling of its surface, beginning to form sulci and gyri (Rakic 1978). This is perhaps why the phenomenon of neuronal cell migration

A. M. Galaburda / Y. Christen (Eds.)
Normal and Abnormal Development of Cortex
© Springer-Verlag Berlin Heidelberg New York 1997

was originally inferred from the observation in histological sections of human embryos by Wilhelm His (1874). However, relatively little progress was made in terms of the underlying cellular and molecular mechanisms of this phenomenon until the 1970s, when a combination of electron microscopic and ^{3}H-thymidine autoradiographic analyses revealed that postmitotic neurons migrate across the intermediate zone by following elongated shafts of non-neural elements called radial glial cells (Rakic 1972). These bipolar cells stretch their processes across the fetal cerebral wall from the onset of corticogenesis, but become most prominent during midgestation, when many of them temporarily stop dividing (Schmechel and Rakic 1979). During the migratory period, cohorts of postmitotic cells originating in the proliferative mosaic follow a radial pathway consisting of single or, more often, multiple glial fibers, which span the developing cerebral wall (Fig. 1). While moving, migrating neurons may be in contact with many axonal and dendritic processes but, nevertheless, remain preferentially attached to glial fibers. This observation suggested a "gliophilic" mode of migration that is mediated by heterotypic adhesion molecules present on neuronal and glial cell surface (Rakic 1985; Rakic et al. 1994). The postmitotic cells that did not obey glial constraints and appeared to move along axonal tracts (eg, red bipolar cell in Fig. 1) were considered "neurophilic" (Rakic 1985). Several initial observations suggested that a gliophilic class of migrating neurons may have the binding affinity for the surface of radial glial cells, as they appear not to follow any other class of cellular processes that are present in the developing cerebral wall (Rakic 1972).

Fig. 1. A three-dimensional illustration of the basic developmental events and types of cell-cell interactions occurring during the early stages of corticogenesis, before formation of the final pattern of cortical connections. This cartoon emphasizes radial migration, a predominant mode of neuronal movement that, in primates, underlies the elaborate columnar organization of the neocortex. After their last division, cohorts of migrating neurons (MN) first traverse the intermediate zone (IZ) and then the subplate zone (SP), where they have an opportunity to interact with "waiting" afferents arriving sequentially from the nucleus basalis and monoamine subcortical centers (NB, MA) from thalamic radiation (TR), and from several ipsilateral and contralateral corticocortical bundles (CC). After the newly generated neurons bypass the earlier generated ones situated in the deep cortical layers, they settle at the interface between the developing cortical plate (CP) and the marginal zone (MZ) and, eventually, form a radial stack of cells that share a common site of origin but are generated at different times. For example, neurons produced between embryonic day E40 and E100 in radial unit 3 follow the same radial glial fascicle and form ontogenetic column 3. Although some cells, presumably neurophilic in the nature of their surface affinities, may detach from the cohort and move laterally, guided by an axonal bundle (eg, horizontally oriented, black cell leaving radial unit 3), most postmitotic cells are gliophilic, eg, they have an affinity for the glial surface and strictly obey constraints imposed by transient radial glial scaffolding (RG). This cellular arrangement preserves the relationships between the proliferative mosaic of the ventricular zone (VZ) and the corresponding protomap within the SP and CP, even though the cortical surface in primates shifts considerably during a massive cerebral growth encountered in mid-gestation (for details see Rakic 1988, 1995 b) ▶

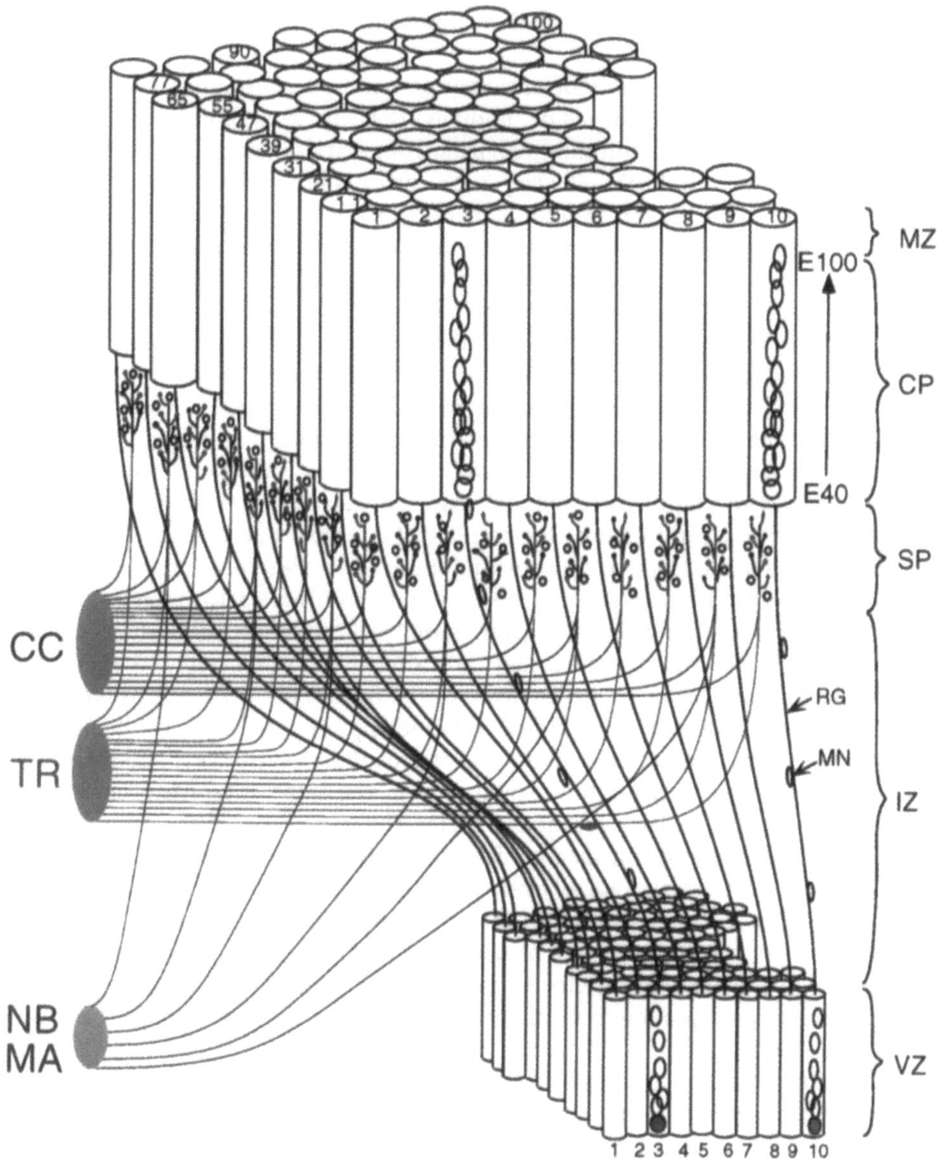
100
90
77
65
55
47
39
31
21
11
1 2 3 4 5 6 7 8 9 10
E100
E40
MZ
CP
SP
IZ
VZ
RG
MN
CC
TR
NB
MA
1 2 3 4 5 6 7 8 9 10

Radial Unit Hypothesis of Cortical Development

The adult neocortex consists of an array of iterative neuronal groups (called, interchangeably, radial columns or modules) that are interconnected in the vertical dimension, share a common extrinsic connectivity and subserve the same function (Mountcastle 1997). The larger the cortex is in a given species, the larger is the number of participating columnar units (Rakic 1978, 1995 a). Although the exact relationship between the functional columns in adult cortex and embryonic columns is not clear, the deployment of migrating neurons into radial columns during cortical development in a given individual and the large increase of cortical surface during evolution by the addition of such columns can be explained in the context of the *radial unit hypothesis* (Rakic 1988). This hypothesis was originally based on the observation that postmitotic cells preserve their positions during movement by remaining attached to a given radial glial fascicle once they initiate their movement (Rakic 1972, 1981). Furthermore, the pattern of distribution of heavily and lightly ^{3}H-thymidine-labeled cells indicated that the columns likely consist of several clones (Rakic 1988). In the past decade the *radial unit hypothesis* has served as a useful working model for research on the cellular and molecular mechanisms involved in normal and abnormal cortical development.

The proposal that cells in a given radial column may be clonally related became possible to test experimentally only after the introduction of the retrieval gene transfer method for *in vivo* analysis of cell lineages in the mammalian brain (Sanes 1989). Although, the clonal relationship of labeled cells using this approach was originally based on the law of probability, cell distribution strongly suggested that most progenitors originating in the same site of the ventricular zone remain radially deployed in the cortex (Luskin et al. 1988). A number of studies in chimeric and transgenic mice have provided more direct evidence that a majority of postmitotic, clonally-related neurons move and remain radially distributed in the cortex (Nakatsuji et al. 1991; Tan and Breen 1993; Soriano et al. 1995; reviewed in Rakic 1995 a). The use of the retroviral gene transfer method in the embryonic primate brain showed that even in the large and highly convoluted cerebrum, radial deployment of many clones is remarkably preserved (Kornack and Rakic 1995).

In addition to those clones that are distributed radially (Luskin et al. 1988), certain populations of clonally related cortical cells do not obey strict radial constraints. Although lateral dispersion of postmitotic neurons was observed in the early Golgi studies (eg, Figure 1 of the report of the Boulder Committee, 1970), this phenomenon has attracted renewed attention after the advent of the retroviral method. Studies in rodents suggested widespread dispersion of some clonally related cortical cells (Walsh and Cepko 1988). In the ferret telencephalon, about 15 % of neurons appear to migrate non-radially (O'Rourke et al. 1992), although it is not clear how many of them actually enter the cerebral cortex. The dispersion is more prominent at later developmental stages, suggesting that these cells may originate from the subventricular zone (J.G. Gleeson and C. Walsh, this volume). It should be emphasized, however, that even those clonally related cells that disperse widely appear to migrate radially along glial fibers situated close to the site of their last mitotic division, a finding that is in full harmony with the radial unit hypothesis (Reid et al. 1995). In primates, it is likely that an even larger proportion of migrating neurons obey the radial constraints imposed by a more elaborate radial glial scaffolding than in rodents (Rakic 1988, 1995 a).

The radial unit hypothesis provides an explanation for a large expansion of cortical surface without a significant increase in thickness that occurred during phylogenetic and ontogenetic development (Rakic 1988). It was proposed that the gene controlling the number of proliferative units at the ventricular surface sets a limit on the size of the cortical surface during individual development, as well as during evolution of the mammalian species (Rakic 1995 b). A relatively small change in the timing of developmental cellular events could have large functional consequences. For example, a minor increase in the length of cell cycles or the number of cell divisions in the ventricular zone can result in a large increase in the number of founder cells that form proliferative units (Rakic 1988). Since proliferation in the ventricular zone initially proceeds exponentially by the prevalence of symmetrical divisions, an additional round of mitotic cycles during this phase doubles the number of proliferative units and, consequently, the number of radial columns (Rakic 1995 b). According to this model, fewer than four extra rounds of cell divisions can account for the 10-fold difference in size of cortical surface between monkey and human. In contrast, the 1000-fold difference between the size of cerebral cortex in mouse and human can be achieved by less than seven extra symmetrical divisions in the ventricular zone before the onset of corticogenesis.

After the number of founder cells that will form individual radial units is set and corticogenesis has begun, many progenitor cells start to divide asymmetrically. Therefore, during this phase of development, an extra round of cell divisions would have a negligible effect on the thickness of the cortex (see Fig. 2 in Rakic 1995 a). According to this model, one can predict that the about two-week-longer duration of corticogenesis in human than in macaque should enlarge the cortical thickness by only 10% to 20%, which is actually observed (Rakic 1995 b). In contrast, even a small delay of the onset of the second phase of corticogenesis results in an order of magnitude larger cortical surface due to the increasing number of proliferative units at the ventricular zone.

Molecular Mechanism of Migratory Pathway Recognition

The universality of neuronal guidance by radial glial scaffolding has been confirmed in a variety of mammalian species, including the human (Sidman and Rakic 1973; Kadhim et al. 1988; Hatten and Mason 1990; Misson et al. 1991; O'Rourke et al. 1995). Initially, based on the inference from the neuron-glial relationship observed *in situ*, it was proposed that a single pair of binding, complementary molecules with gliophilic properties can account for the entire guidance phenomenon of radial migration (Rakic 1981). However, in the last decade, several putative recognition and adhesion molecules have been identified (eg, Cameron and Rakic 1994; Edelman 1983; Hatten and Mason 1986, 1990; Schachner et al. 1985; Anton et al. 1996 b). So far, accumulated evidence suggests it is likely that more than one class of surface molecules must be involved to account for the phenomenon of recognition, adhesion and cessation of neuronal cell migration (Rakic et al. 1994).

An important feature of potential recognition molecules is that they should be expressed selectively and transiently in the leading process of post-mitotic neurons at the surface adjacent to radial glial fibers (Fig. 2). Although the heterotypic recognition mechanism may be of low affinity, these molecules that are an integral part of the membrane can selectively recognize and bind to specific carbohydrate moieties on the adjacent cell. A major advance in this area was identification of the astrotactin, a molecule present on the surface of migrating neurons, blockage of the which curtails neuronal migration *in vitro* (Fishell and Hatten 1991). We recently generated a polyclonal antiserum (D4) and monoclonal antibody (NJPA1) that recognizes a polypeptide forming plasmalemmal junctions situated between migrating neurons and adjacent radial glial fibers (Cameron and Rakic 1994; Cameron et al., submitted for publication). The integrity of these junctional microdomains is maintained in the developing cerebral wall only during the phase of cell migration. In dissociated glial-neuronal cell cultures these microdomains require intact microtubules and can be detected at the sites where the somal region of migrating neurons is in direct contact with elongated glial fibers. An addition of D4 and NJIA1 to the culture medium in imprint and slice preparation of the embryonic cerebral wall leads to withdrawal of the leading process, changes in microtubular organization and, finally, detachment of neurons from radial glial shafts (Anton et al. 1996 a).

After passing between previously generated neurons already settled in the deeper strata of cortical plate, migrating neurons abruptly stop their movement at the border of the marginal zone (MZ in Fig. 2). Actually, the leading process freely enters the marginal zone; it is the movement of the nucleus and surrounding cytoplasm that stops at this barrier (Fig. 2). Only in certain pathological con-

Fig. 2. Model of a proposed cascade of molecular events that take place during migration of postmitotic across developing cerebral wall. Migrating cells extend a leading process (LP) that follows selectively the contours of the radial glial fiber (RF) as it spans the expanding cerebral wall. The cytoskeleton within the LP and trailing process (TP) contain prominent assemblies of microtubules (MT) and actin-like contractile proteins (AC) that are involved in elongation of the LP and translocation of the nucleus (N) and the surrounding cytoplasm. The LP enters the cortical plate (CP) and marginal zone (MZ), but the nucleus stops at the CP/MZ interface (gray area). Various intracellular, membrane-bound and extracellular matrix molecules provide signals or are directly engaged in selection of migratory pathway, rate of cell movement, and finally in the cessation of migration at the CP/MZ borderline. Further explanation in text. Abbreviations: AC, actin-like filaments; AM, homotypic adhesion molecule; CR, Cajal-Retzius cell; CT, catenin; EAA, excitatory amino acids; EF, endfoot of radial glial fiber; Glu, glutamate; Gly, glycine; I, integrine; RM(g) gliophilic recognition molecule; RM(n) neurophilic recognition molecule; Ty, site of increased tyrosine phosphorylation ▶

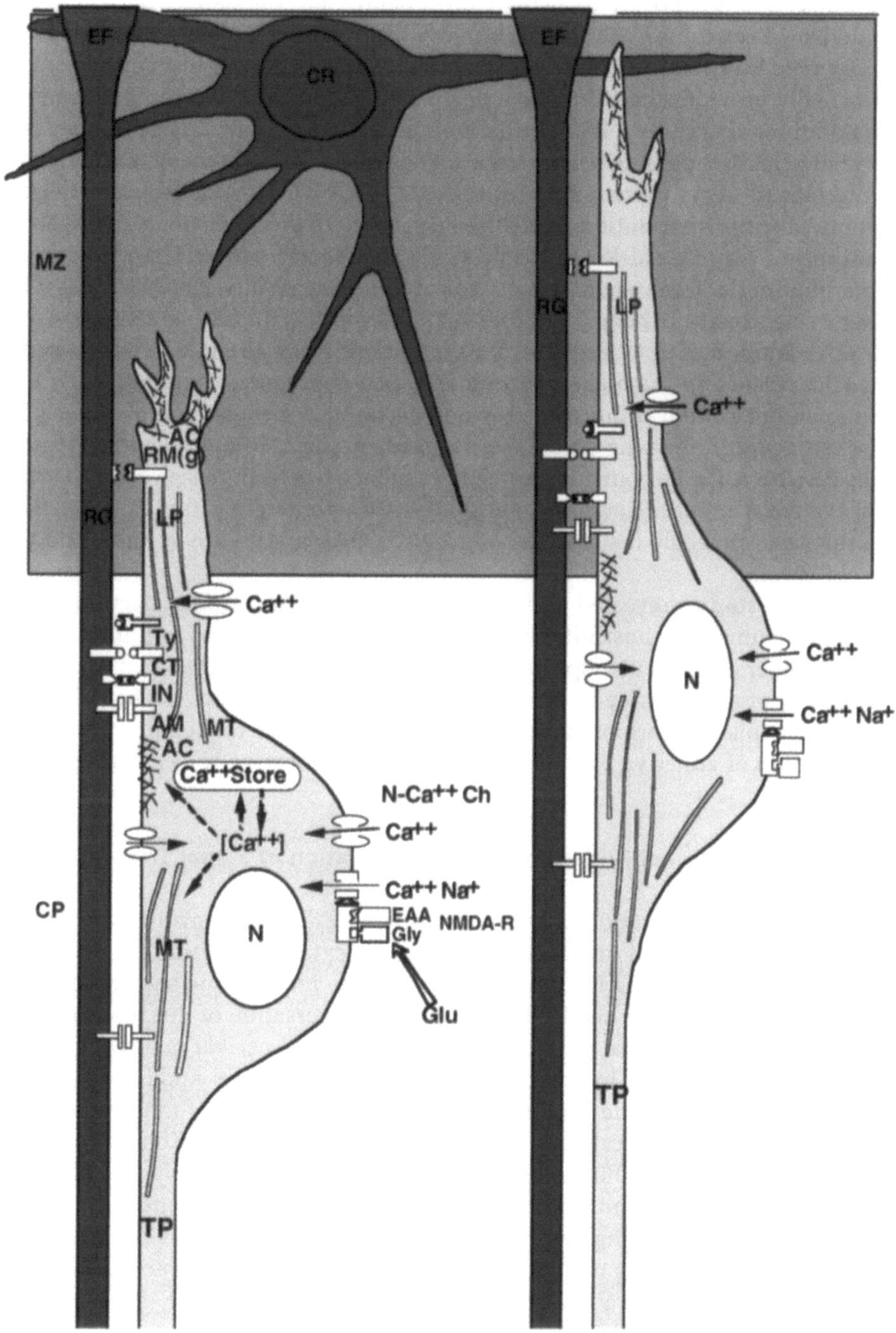
EF
EF
CR
MZ
RG
LP
AC
RM(g)
RG
LP
Ca++
Ty
CT
IN
AM
MT
AC
Ca++Store
N-Ca++ Ch
[Ca++]
Ca++
Ca++ Na+
EAA
Gly
NMDA-R
CP
MT
N
Glu
TP
Ca++
N
Ca++
Ca++ Na+
TP

ditions such as exposure to alcohol, or in neurological mutants such as the reeler and New Zealand mice, do migrating cortical neurons invade the marginal zone (Caviness and Rakic 1978; Sherman et al. 1985). However, the mechanism that normally provides the CP/MZ barrier is not understood. We recently generated an additional antibody, PS2A1, that immunoreacts only with the segment of the radial glial fiber passing through the cortical plate and disappears abruptly at the interface to the marginal zone (Anton et al. 1996 b). This distribution stands in contrast to the distribution of reelin, the product of the *reeler* gene, that is abundant in immature Cajal-Retzius cells and is distributed predominantly within the marginal zone (Ogawa et al. 1995). Our preliminary results, using BrdU as a marker of neuronal birthdays, indicate that postmitotic migrating cells exposed to PS2A1 antibody fail to detach from the glial fibers at the interface between the cortical plate and the marginal zone and, as a consequence, subsequently arriving neurons cannot come at the top and accumulate beneath the previously generated neurons, forming an outside-to-inside gradient of neurogenesis (Anton et al. 1996 b). A similar outside-to-inside sequence has been observed in the cortex of the *reeler* mouse (Caviness and Rakic 1978), although in this case a deficit of a different molecule is likely to be involved (Ogawa et al. 1995; Rakic and Caviness 1996).

These studies revealed that the molecular mechanism for selection of a migratory pathway by postmitotic neurons is more complicated and involves a larger number of molecular species than initially predicted. However, the fast growing body of data obtained from new experimental approaches begins to elucidate the mechanisms by which developmental genes regulate the temporal and spatial expression of adhesive proteins controlling cell allocation to cerebral cortex.

Molecular Control of Initiation, Rate and Cessation of Somal Translocation

Active, directed migration of neurons across long distances involves the unidirectional translocation of the nucleus and somatic cytoplasm within the cytoplasmic cylinder of the growing leading processes. It also requires the signals controlling the cellular machinery involved in initiation and cessation of this movement. The mechanism of the actual physical displacement of cell perikarya has only recently been sought (reviewed in Rakic and Komuro 1995). Voltage- and ligand-gated ion channels situated on the leading process and the cell soma of migrating neurons regulate the influx of calcium ions into migrating neurons (Komuro and Rakic 1996). The influx and fluctuation of free Ca^{++}, in turn, may trigger polymerization of cytoskeletal and contractile proteins essential for translocation of the nucleus and surrounding cytoplasm (Fig. 2). Using live slice and explant preparations of the developing neural tissue to measure the effect of various substances on the rate of cell migration, we found that the voltage-gated, N-type Ca^{++} channel (Komuro and Rakic 1992) and ligand-gated, NMDA channel (Komuro and Rakic 1993) are particularly active. The composition of the receptor/channel complexes on the surface of migrating neurons is different from the composition they have after reaching their final destination (Farrant et al. 1994). These receptor/channels can regulate Ca^{2+} levels in the migrating neurons (Rossi and Slater 1993). We have observed that the amplitude and frequency components of Ca^{2+} fluctuations are positively correlated with the rate of migrating granule cells in cerebellar microexplant cultures (Komuro and Rakic 1996). Moreover, depression of the amplitude and frequency components of Ca^{2+} fluctuations by blockade of calcium influx across the plasma membrane results in a reversible retardation of cell movement. These results indicate that the combination of amplitude and frequency components of intracellular Ca^{2+} fluctuations may provide an intracellular signal controlling the rate of neuronal cell migration.

To explore the cellular machinery underlying translocation of the nucleus during neuronal migration, we determined the polarity of microtubule assemblies situated within the leading and trailing processes of migrating cerebellar granule cells *in situ* (Rakic et al. 1996). Our analysis reveals that the newly assembled, positive ends of the microtubules situated in the leading process are uniformly facing the growing tip, whereas their disintegrating negative ends face the nucleus. In the trailing process, by contrast, microtubule arrays are of mixed polarity. Based on these results, we have proposed that the extension of the leading process may, at least in part, be created by the "push" force of the array of polymers built up in the direction of leading process extension. In contrast, the nucleus and surrounding cytoplasm may be translocated within the membrane envelope because of the orchestrated dissociation of microtubules at their negative ends in front of the nucleus. Such a synchronized disintegration may create a cytoplasmic space devoid of stable cytoskeletal scaffolding, as well as "pull" forces essential for this displacement (Rakic et al. 1996). These "pull" forces are expected to be stronger in the leading process, where the polarity of the microtubule array is uniform, than in the trailing process, where their orientation is mixed. A rapid, coordinated depolymerization of microtubule sheets alternating with a phase of relative stability at their negative ends, as observed during locomotion of non-neuronal cells (Kirshner and Mittchison 1986), may underlie the alternation of movement and stationary periods observed during nuclear displacement in migrating neurons. Our hypothesis that the somatic translocation of the migrating cell may, at least in part, depend on the dynamics of polymerization, and depolymerization of the microtubule protein in the cytoplasm of the leading and trailing processes is consistent with the finding that the slow extension of the leading process of migrating neurons precedes the phase of more rapid nuclear displacement (Hatten and Mason 1990; Komuro and Rakic 1993, 1995). In addition, it is in harmony with the finding that the disruption of the microtubule structure results in the collapse of the migrating cell body and cessation of nuclear translocation (Rivas and Hatten 1995). However, it should be underscored that the proposed role of microtubules in nuclear displacement during neuronal migration does not exclude the synergistic action of actin-like contractile proteins, which may also participate in this event (Rakic 1985).

Signals that induce the onset and cessation of nuclear translocation are unknown, but repolarization of microtubules seems to be an essential prerequisite for this translocation. Caged by the cytoskeletal network, the nucleus could not move without rearrangement of microtubules. The rate of assembling the microtubule polymer depends on the concentration of cytosolic Ca^{2+}, which is delivered through specific voltage- and ligand-gated channels (Fig. 2). In addition, it has been suggested that the integrity and orientation of cytoskeletal proteins may be related to membrane bound cell adhesion molecules (Rakic et al. 1994) and to neuron-glial junctional complexes in migrating neurons (Cameron and Rakic 1994). Identification of molecules that abruptly change their density at the interface of the cortical plate and marginal zone provides promising clues (Ogawa et al. 1995; Anton et al. 1996 a). Likewise, a sudden change in glutamate concentration at the interface to layer I can regulate channels/receptors that provide the signal for cytoskeletal stabilization (Komuro and Rakic 1996). Interruption of the molecular messages in the chain of signals controlling assembly and disassembly of the cytoskeleton in migrating neurons may explain abnormal neuronal placements in certain genetic and acquired brain malformations.

Protomap Hypothesis of Cytoarchitectonic Diversity

A major challenge to developmental neurobiology is how individual and species-specific cytoarchitectonic areas have emerged from the initially seemingly uniform ventricular zone and cortical plate. Both intrinsic and extrinsic factors have been suggested. One attractive hypothesis is that all cortical neurons are equipotential and that laminar and areal differences are induced by extrinsic influences exerted *via* thalamic afferents (Creutzfeldt 1977). Indeed, the structural and functional homogeneity of the cortex emphasized by Lashley and his followers in the middle of this century (eg, Lashley and Clark 1946) appears to be supported by the ostensible similarity in the laminar pattern of cell distribution in various cortical areas and by the acknowledged resemblance of their synaptic circuitry (Eccles 1984; Rakic and Singer 1988; Szentagothai 1987; Mountcastle 1997). This hypothesis is also in harmony with findings that a single-unit response with the typical visual receptive field properties can be elicited from the somatosensory cortex when input from the retina is experimentally rerouted to the somatosensory thalamus (Frost and Metin 1985; Sur et al. 1988). Indeed, there is overwhelming evidence that the development of many features of cortical organization is highly dependent on interaction with thalamic afferents (Rakic 1976 a 1981; Hubel et al. 1977; Shlaggar and O'Leary 1992; Shatz 1996).

However, there is also considerable evidence that the cells generated within the embryonic cerebral wall contain some basic programs of their prospective species-specific cortical organization. To reconcile existing experimental and descriptive data, we formulated a *protomap hypothesis* (Rakic 1988). This hypothesis suggests that the basic pattern of cytoarchitectonic areas emerges through synergistic, interdependent interactions between developmental programs intrinsic to cortical neurons and extrinsic signals supplied by specific input from the subcortical structures. According to this hypothesis, neurons of the embryonic cortical plate – indeed even the proliferative ventricular zone where they originate – set up a primordial species – specific cortical map of areas that preferentially attract appropriate afferents and have a capacity to respond to this input in a specific manner. The prefix *proto* was introduced to emphasize the primordial, provisionary and essentially malleable character of the protomap that is subject to considerable modification by the extrinsic influences exerted at later stages (Rakic 1988).

The initial indication that developmental events in the ventricular zone foreshadow prospective regional differences in the overlaying cerebral mantle comes from ^{3}H-thymidine labeling of dividing cells, which shows that the ventricular region subjacent to area 17 produces more neurons per radial unit (Rakic 1976b) and that the region subjacent to area 17 has a substantially higher mitotic index than the adjacent area 18 (Dehay et al. 1993). Therefore, certain region-specific differences in production of the ventricular zone can be detected even before neurons arrive at the cortex and become exposed to the input from the periphery *via* the thalamic afferents (Kennedy and Dehay 1993; Algan and Rakic 1997). The development of correct topological connections in anophthalmic mice and in early enucleated animals indicates that a basic map can form in the absence of information from the periphery (Kaiserman-Abramoff et al. 1983; Olivaria and Van Sluyters 1984; Rakic 1988). Likewise, the primary visual cortex in the monkey acquires a normal pattern of cytochrome oxidase (Kennedy et al. 1990; Rakic 1988; Kuljis and Rakic 1990) and major neurotransmitter receptors (Rakic and Lidow 1995) in the absence of any retinal input *via* the thalamus from the early embryonic stages. Finally, development of the orientation preference maps in the visual cortex is unaffected by either visual experience or the anatomical rearrangements of the geniculocortical connection into eye-specific domains (Chapman et al. 1996).

The protomap hypothesis recently received support from several reports that show the emergence of area-specific antigens in the cortex before, or independent of, thalamic input (eg, Arimatsu et al. 1992; Ferri and Levitt 1993; Barbe and Levitt 1992; Levitt 1994; Porteus et al. 1992; Buffone et al. 1993). Recent *in vivo* and *in vitro* studies show a high specificity of the growing thalamic axons for the selective regions of the embryonic cerebral cortex (Agmon et al. 1995; Boltz et al. 1992). It was also observed that specific functions can be restored only after transplantation of embryonic cells derived from the same, but not other, areas (Barth and Stanfield 1994). It is likely, therefore, that regulatory genes that are differentially expressed in the cerebral wall could operate in the mammalian telencephalon, but only a few have been tested (eg, Lu et al. 1992; Rubinstein et al. 1994). So far, perhaps the most compelling evidence in favor of the protomap hypothesis has been obtained using transgenic mice combined with transareal transplantation of the embryonic cerebral wall to other regions (Cohen-Tannoudji et al. 1994). This study indicates that certain genes are expressed selectively in the somatosensory cortex without, and irrespective of, the type of afferents that neurons destined for this region receive from the thalamus.

It should be underscored that although the embryonic cerebral wall exhibits gradients of morphoregulatory molecules as well as more specific area-specific molecular differences, the protomap within the embryonic cerebrum provides only a set of species-specific genetic instructions and biological constraints. The precise position of interareal borders, the overall size of each cytoarchitectonic area, and the details of their cellular and synaptic characteristics in the adult cerebral cortex are achieved through a cascade of reciprocal interactions between cortical neurons and cues they receive from afferents arriving from a variety of extracortical sources (Rakic 1988). Such afferents may serve to coordinate and adjust the ratio of various cell classes with the subcortical structures, as has been shown in the primary visual system (Messirel et al. 1997). For example, the experimental manipulation of geniculocortical input to the visual cortex indicates that the size of a given cytoarchitectonic area can be regulated by afferents originating from subcortical structures (Rakic et al. 1991; Kennedy and Dehay 1993; Rakic and Lidow 1995). This extracortical control can be exerted on the rate of cell elimination as well as on cell production in the ventricular zone (H. Kennedy and C. Dehay, this volume). In summary, the concept of the cortical protomap includes the role of both intrinsic and extrinsic determinants in shaping the final pattern and relative size of the cytoarchitectonic areas.

Relevance of Neuronal Migration to Cortical Malformation

Collectively, accumulated evidence suggests that specific communications between heterogeneous classes of cells play a major role in establishing their position, identity, and selection of migratory pathways before they arrive at their final destination in the cortex and form synaptic connections. Thus, multiple causes can produce similar migratory deficits and diverse forms of cortical malformations can be caused by the same molecular events. For example, the process of migration is highly sensitive to various physical (eg, ionizing radiation, heat), chemical (eg, various drugs, excessive use of alcohol), and biological (some neurotropic viruses) agents, as well as to genetic mutations, and is thought to be the major cause of various gross brain malformations, as well as of more subtle abnormalities in neuronal positioning and in the pattern of synaptic circuits.

The most frequent migratory abnormalities are revealed in the mature brain by the presence of solitary ectopic neurons that have failed to reach their target and are scattered within the white matter. The cause for such misplacements may be exposure to various exogenous agents such as low doses of irradiation or drugs that interfere with signaling pathways (Lidwo 1995; Algan and Rakic 1997). Despite their small number, such ectopic cells may be a major source of epileptic discharge *via* abnormal synaptic connections with the overlaying cortex. However, more massive malformations, known by descriptive terms such as microencephaly, schizencephaly, lissencephaly, macrogyria, polymicrogyria or "double cortex", may be fully or in part consequential to defective migration and abnormal settling of nerve cells during the formative stages of the human neocortex. Advances made in developmental neurobiology allow some new interpretations of these malformations.

According to the radial unit hypothesis, the pathogenesis of major cortical malformations can be classified into two major categories. The first category comprises malformations in which the number of radial units in the cortex is reduced, whereas the number of neurons within each ontogenetic column remains relatively normal. It can be expected that defects in this category result from an early occurring event that alters the number of proliferative units at the time they are being formed in humans within the first six weeks of gestation. Once the number of proliferative units in the ventricular zone is established, albeit in fewer numbers, each unit can produce a normal, or even greater, number of neurons that become crowded in the diminished cerebral vesicle. It could be expected that the cortex would have a smaller surface area in spite of a normal or enlarged thickness and the presence of massive neuronal ectopias in the white matter. Identification of defective genes in humans that may cause a smaller cerebral surface with an absence of convolutions, presumably by disrupting neuronal migration, opens the prospect of unraveling the pathogenesis of such types of malformations at the molecular level (Reiner et al. 1993).

The second category consists of malformations in which the initial formation of radial units is not affected, whereas the number of neurons within radial columns in the cortex is reduced. The defect in this category should begin after the sixth fetal week, when the normal complement of proliferative units in the human cerebrum has already been established. Such malformations can be caused by interference with cell proliferation *via* intrinsic (genetic) or extrinsic (irradiation or viral infections) factors. Diminished production of neurons in the proliferative units results in fewer neurons within ontogenetic columns, and the cortex is therefore thinner. It should be recognized that in most cases cortical malformations may have features of only one or the other category, but in practice most show a mixture of both.

The proposed classification of cortical malformations suggests possible developmental mechanisms by separating defects of unit formation from defects of ontogenetic column formation. In support of the radial unit hypothesis and the concept of a cortical protomap, experimental manipulations and neuropathological data reveal that each step (formation of proliferative units, formation of ontogenetic columns, and formation of cytoarchitectonic areas) can be separately affected by genetic defects or by extrinsic factors. This general model of the differential effect of early and late disturbance of corticogenesis on the surface and thickness of the cerebral cortex has also been tested by experimental manipulation of corticogenesis in primates. For example, targeted X-irradiation of the fetal monkey brain during the phase of corticogenesis (E40–E100) eliminates neurons from specific layers (Algan and Rakic 1996), as expected from their laminar assignment (McConnell 1988). However, although in such cases the width of the cortex is diminished, the surface area of the cortex remains unchanged. In contrast, in animals irradiated with low doses of X-ray before the onset of corticogenesis (E33–40), area 17 displayed only slight changes in cortical thickness, cell density and area-specific cytoarchitectonic features, whereas the total surface devoted to area 17 was significantly diminished (Algan and Rakic 1997). In contrast, in animals irradiated with low doses during the period of corticogenesis, the surface of area 17 was not significantly altered, although the cortical layers generated at the time of irradiation had a significantly lower number of cells per radial unit. This interpretation has also been supported by the mode of cortical development in the trisomy 16 mouse, which is considered to be an animal model of Down syndrome (Haydar et al. 1996). In this malformation slow tangential expansion of neuroepithelium in Ts16 mice resulted in a reduction in the final telencephalic size that can be predicted by a reduction of radial units in early telencephalic development.

The number of neurons in radial columns could also be affected by excessive proliferation, diminished apoptosis, and a failure of neuronal migration. In such cases, supernumerary neurons may fail to reach the cortex and often survive in ectopic positions within the white matter. An instructive example in this category is the abnormal development of cerebral cortex in transgenic mouse, which was made using homologous recombination deficient in CPP32 protease that is essential for a normal process of programmed cell death (apoptosis; Kuida et al. 1996). In homozygous mice supernumerary cells were consistently located between the ventricular zone and the cerebral cortex in a remarkably similar fashion to the deployment in the "double cortex" (Palmini et al. 1991; J. G. Gleeson and C. Walsh, this volume).

The differential effect on the surface and thickness of the cortex can be observed in a variety of other malformations, although the pathogenesis in most cases still remains obscure (eg, Volpe 1987). Understanding the cellular and molecular mechanisms of neuronal proliferation and migration may help to explain the pathogenesis of the phenotype of previously inexplicable genetic and acquired conditions observed in childhood epilepsy, mental retardation, schizophrenia or developmental dyslexia (Aicardi 1991; Palmini et al. 1991; Volpe 1987; Galaburda et al. 1985; Bloom 1993).

Acknowledgments

This work was supported over the years by grants from the US Public Health Service. I am indebted to the present and past members of my laboratory for their contributions and incisive discussion on this subject.

References

Agmon AA, Yang LT, Jones GE, Dowd DK (1995) Topological precision in the thalamic projection to the neonatal mouse. J Neurosci 13:5365–5382

Aicardi J (1991) The agyria-pachigyria complex: A spectrum of cortical malformations. Brain Dev 13:1–8

Algan O, Rakic P (1997) Radiation-induced area- and lamina-specific deletion of neurons in the primate visual cortex. J Comp Neurol 381:335–352

Anton SA, Cameron RS, Rakic P (1996a) Role of neuro-glial junctional proteins in the maintenance and termination of neuronal migration across the embryonic cerebral wall. J Neurosci 16:2283-2293

Anton ES, Matthew WD, Rakic P (1996b) A regionally distributed radial glial antigen: A candidate for signaling an end to neuronal migration. Soc Neurosci Abstr 22:1206

Arimatsu Y, Miyamoto M, Nihonmatsu I, Hirata K, Urataini Y, Hatanka Y, Takiguchi-Hoyash K (1992) Early regional specification for a molecular neuronal phenotype in the rat neocortex. Proc Nat Acad Sci USA 89:8879-8883

Barbe MF, Levitt P (1992) Attraction of specific thalamic input by cerebral grafts depends on the molecular identity of the implant. Proc Nat Acad Sci USA 89:3706-3710

Barth TM, Stanfield BB (1994) Homotopic, but not heterotopic, fetal cortical transplants can result in functional sparing following neonatal damage to the frontal cortex in rat. Cereb Cortex 4:271-278

Bloom RE (1993) Advancing a neurodevelopmental origin of schizophrenia. Arch Gen Psychiatry 50:224-227

Boltz J, Novak N, Staiger V (1992) Formation of specific afferent connections in organotypic slice cultures from rat visual cortex cocultured with lateral geniculate nucleus. J Neurosci 12:3054-3070

Boulder Committee (1970) Embryonic vertebrate central nervous system. Revised terminology. Anat Rec 166:257-261

Buffone A, Kim HJ, Puelles L, Proteus MH, Frohman MA, Martin GR, Rubinstein JLR (1993) The mouse DLX-2 (Tes-1) Gbx-2 and Wnt-3 in the embryonic day 12.5 mouse forebrain defines potential transverse and longitudinal segmental boundaries. Mech Dev 40:129-140

Cameron RS, Rakic P (1994) Polypeptides that comprise the plasmalemal microdomain between migrating neuronal and glial cells. J Neurosci 14:3139-3155

Caviness VS Jr, Rakic P (1978) Mechanisms of cortical development: A view from mutations in mice. Ann Rev Neurosci 1:297-326

Chapman B, Stryker MP, Bonhoeffer T (1996) Development of orientation preference maps in ferret primary visual cortex J Neurosci 16:6443-6453

Cohen-Tannoudji M, Babinet C, Wassef M (1994) Early intrinsic regional specification of the mouse somatosensory cortex. Nature 368:460-463

Creutzfeldt OD (1977) Generality of the functional structure of the neocortex. Naturwissenschaften 64:507-517

Dehay C, Giroud P, Berland M, Smart I, Kennedy H (1993) Modulation of the cell cycle contributes to the parcellation of the primate visual cortex. Nature 366:464-466

Eccles JC (1984) The cerebral neocortex: A theory of its operation. In: Jones EG, Peters A (eds) Cerebral cortex. Vol 2. Plenum Press, New York, pp 1-36

Edelman GM (1983) Cell adhesion molecules. Science 219:450-457

Farrant M, Feldmeyer D, Takahashi T, Cull-Candy SG (1994) NMDA-receptor channel diversity in the developing cerebellum. Nature 368:335-339

Ferri RT, Levitt P (1993) Cerebral cortical progenitors are fated to produce region-specific neuronal populations. Cereb Cortex 3:187-198

Fishell G, Hatten ME (1991) Astrotactin provides a receptor system for CNS neuronal migration. Development 113:755-765

Frost DO, Metin C (1985) Introduction of functional retinal projections to the somatosensory system. Nature 317:162-164

Galaburda AM, Sherman GF, Rosen GD, Aboitiz F, Geschwind N (1985) Developmental dyslexia: Four consecutive patients with cortical abnormalities. Ann Neurol 18:222-223

Hatten ME, Mason CA (1986) Neuron-astroglia interactions *in vitro* and *in vivo*. Trends Neurosci 9:168-174

Hatten ME, Mason CA (1990) Mechanism of glial-guided neuronal migration *in vitro* and *in vivo*. Experientia 46:907-916

Haydar TF, Blue EB, Molliver ME, Krueger BK, Yarowski PJ (1996) Consequences of Trisomy 16 for mouse brain development: corticogenesis in a model of Down syndrome. J Neurosci 16:6175-6182

His W (1874) Unsere Körperform und das Physiologische Problem innerer Entstehung. Engelman, Leipzig

Hubel DH, Wiesel TN, LeVay S (1977) Plasticity of ocular dominance columns in monkey striate cortex. Phil Trans Roy Soc Lond Serv B 278:377–409

Kadhim HJ, Gadisseux J-F, Evrard P (1988) Topographical and cytological evolution of the glial phase during prenatal development of the human brain: Histochemical and electron microscopic study. J Neuropathol Exp Neurol 47:166–188

Kaiserman-Abramov I, Graybiel A, Nauta WH (1983) Thalamic projection to area 17 in a congenitally anophthalmic mouse strain. Neuroscience 5:41–52

Kennedy H, Dehay C (1993) Cortical specification of mice and men. Cereb Cortex 3:171–186

Kennedy H, Dehay C, Horsburgh G (1990) Striate cortex periodicity. Nature 384:494

Kirschner M, Mitchison T (1986) Beyond self-assembly: From microtubules to morphogenesis. Cell 45:329–342

Komuro H, Rakic P (1992) Specific role of N-type calcium channels in neuronal migration. Science 257:806–809

Komuro H, Rakic P (1993) Modulation of neuronal migration by NMDA receptors. Science 260:95–97

Komuro H, Rakic P (1995) Dynamic of granule cell migration: a confocal microscopic study in acute cerebellar slice preparations. J Neurosci 15:1110–1120

Komuro H, Rakic P (1996) Intracellular Ca^{2+} fluctuations modulate the rate of neuronal migration. Neuron 17:275–285

Kornack DR, Rakic P (1995) Radial and horizontal deployment of clonally related cells in the primate neocortex: Relationship to distinct mitotic lineages. Neuron 15:311–321

Kuida K, Zheng TS, Na S, Kuang C, Yang D, Karasuyama H, Rakic P, Flavell R (1996) Decreased apoptosis in the brain and premature lethality in CPP32-deficient mice. Nature 384:368–372

Kuljis RO, Rakic P (1990) Hypercolumns in the monkey visual cortex can develop in the absence of cues from photoreceptors. Proc Natl Acad Sci USA 87:5303–5306

Lashley KS, Clark G (1946) The cytoarchitecture of the cerebral cortex of Ateles: A critical examination of cytoarchitectonic studies. J Comp Neurol 85:223–305

Levitt P (1994) Experimental approaches that reveal principles of cerebral cortical development. In: Gazzaniga M (ed) cognitive neurosciences. MIT Press, Cambridge MA, pp 147–163

Lidow MS (1995) Prenatal cocaine exposure adversely affects the development of the primate cerebral cortex. Synapse 21:332–341

Lu S, Bogorad LD, Murtha MT, Ruddle F (1992) Expression pattern of a marine homeobox gene, Dbx, displays extreme spatial restriction in embryonic forebrain and spinal cord. Proc Natl Acad Sci USA 89:8053–8057

Luskin MB, Pearlman AL, Sanes JR (1988) Cell lineage in the cerebral cortex of the mouse studied *in vivo* and *in vitro* with a recombinant retrovirus. Neuron 1:635–647

McConnell SK (1988) Development and decision-making in the mammalian cerebral cortex. Brain Res Rev 13:1–23

Meissirel C, Wikler KC, Chalupa LM, and Rakic P (1997) Early divergence of M and P visual subsystems in the embryonic primate brain. Proc Nat Acad Sci (USA) 94:5900–5905

Misson JP, Austin CP, Takahashi T, Cepko CL, Caviness VS (1991) The alignment of migrating neural cells in relation to the murine neopallial radial glial fiber system. Cereb Cortex 1: 221–229

Mountcastle VB (1997) The columnar organization of the neocortex. Brain 120:701–722

Nakatsuji M, Kadokawa Y, Suemori H (1991) Radial columnar patches in the chimeric cerebral cortex visualized by use of mouse embryonic stem cells expressing β-galactosidase. Dev Growth Differ 33:571–578

Ogawa M, Miyata T, Nakajima K, Yagyu K, Selke M, Ikenaka K, Yamamoto H, Mikoshiba K (1995) The *reeler* gene-associated antigen on Cajal-Retzius neurons is a crucial molecule for laminar organization of cortical neurons. Neuron 14:1–20

O'Leary DM, Koestler SE (1993) Development of projection neuron types, axon pathways and patterned connections of the mammalian cortex. Neuron 10:991–1006

Olivaria J, Van Sluyters RC (1984) Callosal connections of the posterior neocortex in normal-eyed, congenitally anophthalmic and neonatally enucleated mice. J Comp Neurol 230:249–268

O'Rourke NA, Dailey ME, Smith SJ, McConnell SK (1992) Diverse migratory pathways in the developing cerebral cortex. Science 258:299–302
O'Rourke NA, Dulivan NA, Smith DP, Kazanowski SE, Jacobs AA, McConnell SK (1995) Tangential migration in developing cerebral cortex. Development 1212:2165–2176
Palmini A, Andermann F, Aicardi J, Dulac O, Chavis F, Ponsot G, Pinard JM, Goutieres, Livingston J, Tampieri D, Andermann E, Robitaille Y (1991) Diffuse cortical dysplasia, or the "double cortex" syndrome. Neurology 41:1656–1662
Parnavelas JG, Barfield JA, Franke E, Luskin MB (1991) Separate progenitor cells give rise to pyramidal and nonpyramidal neurons in the rat telencephalon. Cereb Cortex 1:463–491
Porteus MH, Brice EJ, Buffone A, Usdin TB, Ciaranello RD, Rubinstein JR (1992) Isolation and characterization of a library of cDNA clones that are preferentially expressed in the embryonic telencephalon. Mol Brain Res 12:7–22
Rakic P (1972) Mode of cell migration to the superficial layers of fetal monkey neocortex. J Comp Neurol 145:61–84
Rakic P (1974) neurons in the monkey visual cortex: Systematic relation between time of origin and eventual disposition. Science 183:425–427
Rakic P (1976 a) Prenatal genesis of connections subserving ocular dominance in the rhesus monkey. Nature 261:467–471
Rakic P (1976 b) Differences in the time of origin and in eventual distribution of neurons in areas 17 and 18 of the visual cortex in the rhesus monkey. Exp Brain Res Suppl 1:244–248
Rakic P (1978) Neuronal migration and contact interaction in primate telencephalon. Postgrad Med J 54:25–40
Rakic P (1981) Neuron-glial interaction during brain development. Trends Neurosci 4:184–187
Rakic P (1985) Contact regulation of neuronal migration. In: Edelman GM, Thiery J-P (eds) The cell in contact: adhesions and junctions as morphogenetic determinants. Wiley and Sons, New York, pp 67–91
Rakic P (1988) Specification of cerebral cortical areas. Science 241:170–176
Rakic P (1995 a) Radial versus tangential migration of neuronal clones in the developing cerebral cortex. Proc Nat Acad Sci USA 92:11323–11327
Rakic P (1995 b) A small step for the cell – a giant leap for mankind: a hypothesis of neocortical expansion during evolution. Trends Neurosci 18:383–388
Rakic P, Caviness VS (1995) Cortical development: view from neurological mutants two decades later. Neuron 14:1101–1104
Rakic P, Singer W (eds) (1988) Neurobiology of the Neocortex. Wiley and Sons. New York, NY
Rakic P, Lidow MS (1995) Distribution and density of neurotransmitter receptors in the absence of retinal input from early embryonic stages. J Neurosci 15:2561–2574
Rakic P, Komuro H (1995) The role of receptor-channel activity in neuronal cell migration. J Neurobiol 26:299–315
Rakic P, Kornack DR (1993) Constraints on neurogenesis in adult primate brain: an evolutionary advantage? In: Cuello AC (ed) Neuronal cell death and repair. Restorative Neurology 6. Amsterdam, Elsevier, pp 257–266
Rakic P, Cameron RS, Komuro H (1994) Recognition, adhesion, transmembrane signaling, and cell motility in guided neuronal migration. Curr Opin Neurobiol 4:63–69
Rakic P, Knyihar-Csillik E, Csillik B (1996) Polarity of microtubule assembly during neuronal migration. Proc Natl Acad Sci USA 93:9218–9222
Rakic P, Stensaas LJ, Sayre EP (1974) Computer-aided three-dimensional reconstruction and quantitative analysis of cells from serial electron microscopic montages of fetal monkey brain. Nature 250:31–34
Rakic P, Suner I, Williams RW (1991) Novel cytoarchitectonic areas induced experimentally within primate striate cortex. Proc Natl Acad Sci USA 88:2083–2987
Reid C, Liang I, Walsh C (1995) Systematic widespread clonal organization in cerebral cortex. Neuron 15:299–310

Reiner O, Carrozzo R, Shen Y, Wehnert M, Faustinella F, Dobyns WB, Caskey CT, Ledbetter DH (1993) Isolation of a Miller-Dieker lissencephaly gene containing G protein b-subunit-like repeats. Nature 364:717–721

Rivas RJ, Hatten MB (1995) Motility and cytoskeletal organization of migrating cerebellar granule neurons. J Neurosci 15:981–989

Rossi D, Slater TN (1993) The developmental onset of NMDA receptor channel activity during neuronal migration. Neuropharmacology 32:1239–1248

Rubinstein JLR, Martinez S, Shimamura K, Puelles L (1994) The embryonic vertebrate forebrain: The prosomeric model. Science 266:578–580

Sanes JR (1989) Analyzing cell lineages with a recombinant retrovirus. Trends Neurosci 12:21–28

Schachner M, Faissner A, Fischer G (1985) Functional and structural aspects of the cell surface in mammalian nervous system development. In: Edelman GM, Gall WE, Thiery JP (eds) The cell in contact: adhesions and junctions as morphogenetic determinants. John Wiley & Sons, New York

Schlagger BL, O'Leary DDM (1992) Potential of visual cortex to develop an array of functional units unique to somatosensory cortex. Science 252:1556–1560

Schmechel DE, Rakic P (1979) Arrested proliferation of radial glial cells during midgestation in rhesus monkey. Nature 227:303–305

Shatz CJ (1996) Emergence of order in visual system development. Proc Natl Acad Sci USA 93:602–608

Sherman GF, Galaburda AM, Geeschwind N (1985) Cortical anomalies in brains of New Zealand mice: a neuropathologic model of dyslexia? Proc Natl Acad Sci USA 82:8072–8074

Sidman RL, Rakic P (1973) Neuronal migration with special reference to developing human brain: A review. Brain Res 62:1–35

Sidman RL, Rakic P (1982) Development of the human central nervous system. In: Haymaker W, Adams RD (eds) Histology and histopathology of the nervous system. Springfield, IL, Charles C Thomas, pp 3–145

Soriano E, Dumesnil N, Auladell C, Cohen-Tannoudji M, Sotelo C (1995) Molecular heterogeneity of progenitors and radial migration in the developing cerebral cortex revealed by transgenic expression. Proc Natl Acad Sci USA, 92:11676–11680

Sur M, Garraghty PE, Roe AW (1988) Experimentally induced visual projections in auditory thalamus and cortex. Science 242:1434–1441

Szentagothai J (1987) The neuronal network of the cerebral cortex: A functional interpretation. Prog Brain Res 201:219–248

Tan S-S, Breen SJ (1993) Radial mosaicism and tangential cell dispersion both contribute to mouse neocortical development. Nature 362:638–640

Volpe JJ (1987) Neurology of the newborn. Ed. 2. Philadelphia, WB Saunders

Walsh C, Cepko CL (1988) Clonally related cortical cells show several migration patterns. Science 241:1342–1345

Exuberant Growth, Specificity, and Selection in the Differentiation of Cortical axons

G. M. Innocenti and L. Tettoni

Abstract

Callosal connections, a model for cortical connections, are analyzed in their adult organization and development from the point of view of axonal phenotype. The phenotype of adult callosal axons implements multidimensional maps, and in addition, transformations in the temporal domain. Callosal connections develop through several stages of axonal differentiation. Each stage results from the interplay of intrinsic axonal properties with local cues external to the axon. Axonal properties include the apparent commitment of the axon to a specific connectional phenotype and specific growth modes. The latter lead to overproduction of axonal structures, i.e., of axonal trunks, branches and synapses. External cues validate the differentiation of the axon, select the axonal structures that will be maintained and trigger the transition into the next stage of differentiation. External cues include experience-driven activity acting in the phase of synaptic production and therefore, potentially, having a constructive role in axonal differentiation.

Introduction

Current views on the development of neural connections in vertebrates are profoundly influenced by the results of studies conducted on "simple" neural systems. In this context, "simple" neural systems are characterized by orderly, point-to-point connections and/or by one-to-one relations between the pre- and post-synaptic elements. In vertebrates, a typical example is the orderly mapping of the two-dimensional retinal coordinates on the tectum. Studies in this system have led to the notion of target-aimed axonal growth which is explicit in Sperry's chemoaffinity theory of connectional development (Sperry 1963; discussed in Innocenti 1988) and implicit in several modern versions of it. Briefly, according to these theories, the axons are intrinsically different from each other and can recognize specific cues that guide them precisely towards their targets. These cues could be diffusible chemoattractants released from the target or membrane-bound molecules in the environment of the growing axons. The origin of point-to-point maps could be explained on the basis of orthogonal gradients of chemical markers that inhibit or promote axonal growth. In the simple cases, two such

A. M. Galaburda / Y. Christen (Eds.)
Normal and Abnormal Development of Cortex
© Springer-Verlag Berlin Heidelberg New York 1997

markers distributed in orthogonal gradients, at the target, can generate a map. Each point in the map is characterized by a unique concentration of the markers and thus identifies a location to which an axon is specifically attracted (reviewed in Tessier-Lavigne and Goodman 1996; Rétaux and Harris 1996).

This original view of the development of neural connections has been complicated by evidence that the process of axonal growth is, in fact, a sequence of different operations, (axonal elongation, target recognition, etc.) in which different factors are sequentially involved. However, the notion of axonal specificity and target-aimed axonal growth has been strengthened by the discovery of molecules that either promote or inhibit axonal growth (see Goodman and Shatz 1993; Tessier-Lavigne and Goodman 1996; Rétaux and Harris 1996).

As expressed above, the concepts of axonal specificity and target-aimed growth appear insufficient to explain the development of cortical connections. The main reason is that, in cerebral cortex, no simple connectional maps exist. In general, cortical maps have multiple dimensions and obey a hierarchy of mapping rules. Thus, the primary sensory areas do contain homeomorphic representations of the sensory peripheries, similar to the retinal maps found on the tectum. However, within these maps, other variables are mapped as well, such as laterality of sensory organs (eye or ear), sensory submodalities, orientation and direction of stimulus, etc. Complex connectional maps burden theories of the development of connections. Indeed, one or two gradients of molecular markers can no longer account for the growth of individual axons towards their targets. Consequently, if axonal growth has to be precisely target-aimed, the guidance and target recognition mechanisms need to be considerably more complicated.

In apparent opposition to the axonal specificity and/or the target-aimed axonal growth concepts (Changeux and Danchin 1976; Easter et al. 1985; Edelman 1987; Innocenti 1988; O'Leary 1992), a large body of experimental data documents the formation of exuberant projections/connections in the developing nervous system of several species, in particular in cerebral cortex (reviewed in Innocenti 1991). Exuberant projections consist of axons or axonal branches, most frequently identified by retrograde axonal tracers, that are deleted in the course of maturation. They often interconnect brain regions that are no longer connected in the adult. Quantitative estimates of axonal (reviewed in Innocenti 1991) and synaptic (reviewed in Rakic et al. 1994) overproduction in development were provided by electron-microscopic investigations. Thus, in both cat and monkey, about 70% of the juvenile axons interconnecting the cerebral hemispheres through the corpus callosum are eliminated. Regressive events of this order of magnitude are too large to be considered errors in axonal growth.

Instead, the formation of exuberant projections implies the existence of mechanisms operating by selection. The fact that the development of cortical connections can be modified by neural activity (reviewed in Goodman and Shatz 1993; Cramer and Sur 1995) implies that the selection can operate on criteria other than the molecular specificity of the individual neurons (Singer 1995).

Here we review the development of callosal connections among visual areas of the cat, a model for the development of cortical connectivity. The main emphasis will be on axonal phenotype. The concept of axonal phenotype includes substantially more than the orderly topographic point-to-point mapping usually credited to neural connectivity. Indeed, it refers to an open-ended set of axonal features, some of which implement specific operations that are computational in nature. It will be argued that the development of neural connections consists, to a large extent, of a process of conditional axonal differentiation. Axons differentiate through stages at which their individual specificities unfold. Each stage requires validation by environmental cues that also trigger the next stage of differentiation. Hypotheses about the mode of operation of these cues can solve the conundrum of specific vs. exuberant axonal growth. We propose that these cues operate through selection rather than instruction and that they canalize, rather than direct, axonal growth.

Cortico-Cortical Axons Implement Multidimensional Mapping Rules

Visual callosal connections are organized according to topographical principles, that were either discovered or validated by studies using retrogradely transported axonal tracers (reviewed in Innocenti 1986; Innocenti and Bressoud 1997). First, each cortical area is connected in an orderly way with the corresponding area (homotopic callosal connections), but also with a characteristic set of other areas in the contralateral hemisphere (heterotopic callosal connections). The connections are usually reciprocal. Second, cortical connections are unevenly distributed across an area. In the visual system, the callosal connections of primary visual areas are restricted to the border between areas 17 and 18. They are roughly point-to-point along this border, which corresponds to the vertical meridian of the visual field. The border between areas 17 and 18, however, projects not only to the contralateral homologous region but also to areas 19, 21a and to the lateral suprasylvian areas (PMLS and PLLS; Segraves and Rosenquist 1982).

Considerable refinement of the concepts mentioned above came when individual callosal axons were labeled with anterogradely transported byocitin and reconstructed three dimensionally from serial sections (Houzel et al. 1994; Innocenti et al. 1994; Innocenti and Bressoud 1997 and unpublished). Indeed, it was found that individual axons implement mapping rules beyond those mentioned above.

First, several types of axons can be identified, based on their pattern of area-to-area projection (Figs. 1 and 2). The majority of the axons project only to the contralateral 17/18 region. Others project to one or two of the following extrastriate areas: 19, 21a, PMLS, or PLLS. Thus, in general, different neuronal populations are involved in the projection from the primary visual areas to other areas. However, individual axons can branch to several extrastriate areas.

Second, even when restricted to one area, individual axons can distribute terninals over a territory of several mm^2. The degree of divergence varies with the

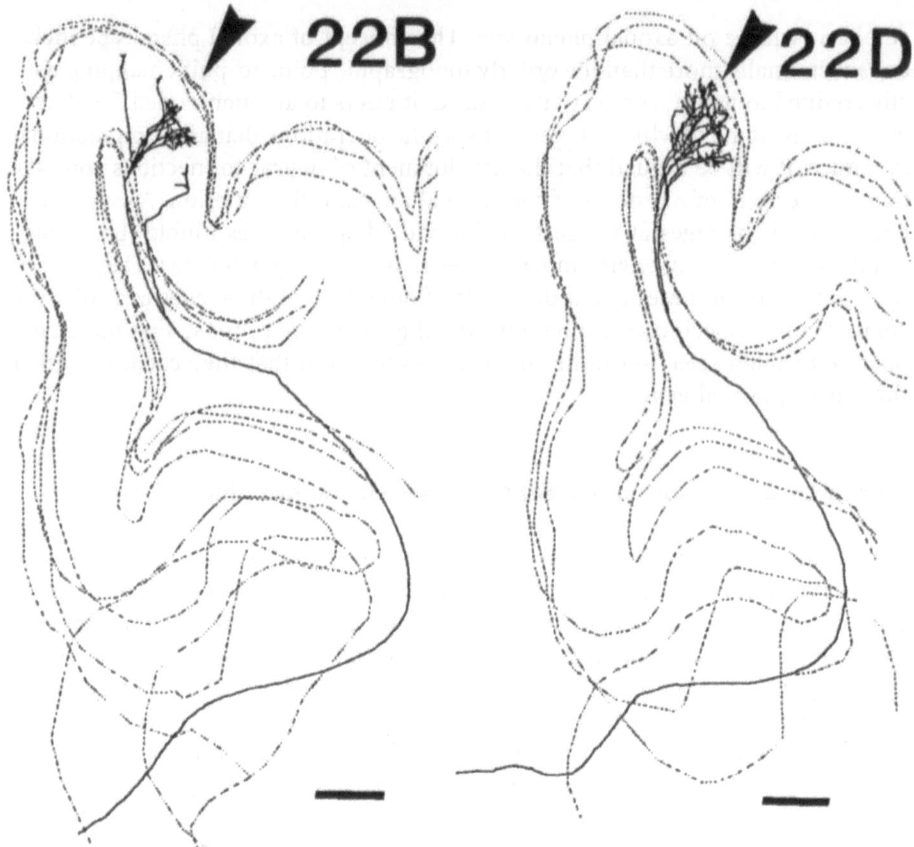

Fig. 1. Computer-aided 3-D reconstructions of individual callosal axons from areas 17 and 18, labeled by anterogradely transported biocytin and terminating near the border between visual areas 17 and 18 (arrowhead) of an adult cat. Code numbers identify experiment; letters identify the individual axons. The outlines of a few of the serial sections used for the reconstruction are also shown; pial surface and bottom of the gray matter are indicated by interrupted lines. Dorsal is upwards, medial to the left. Calibration bars are 500 μm. (From Houzel et al. 1994, modified)

target area. It also varies among axons to the same areas, possibly in relation to the local graininess of the retinotopic map.

Third, individual axons terminate with distinct clusters of boutons. The volume containing a cluster is called a "terminal column." As discussed previously (Houzel et al. 1994), terminal columns presumably correspond to orientation columns, recognizing the same stimulus orientation as the cell body to which the axon belongs. In several instances, though, the diameter of a terminal column is inferior to that of an orientation column. Thus, only part of an orientation column might receive callosal axons.

Fourth, callosal axons do not usually distribute to all layers, although in some cases they do. The laminar distribution can vary at the different terminal columns of one axon, but the majority of boutons are, invariably, in the supragra-

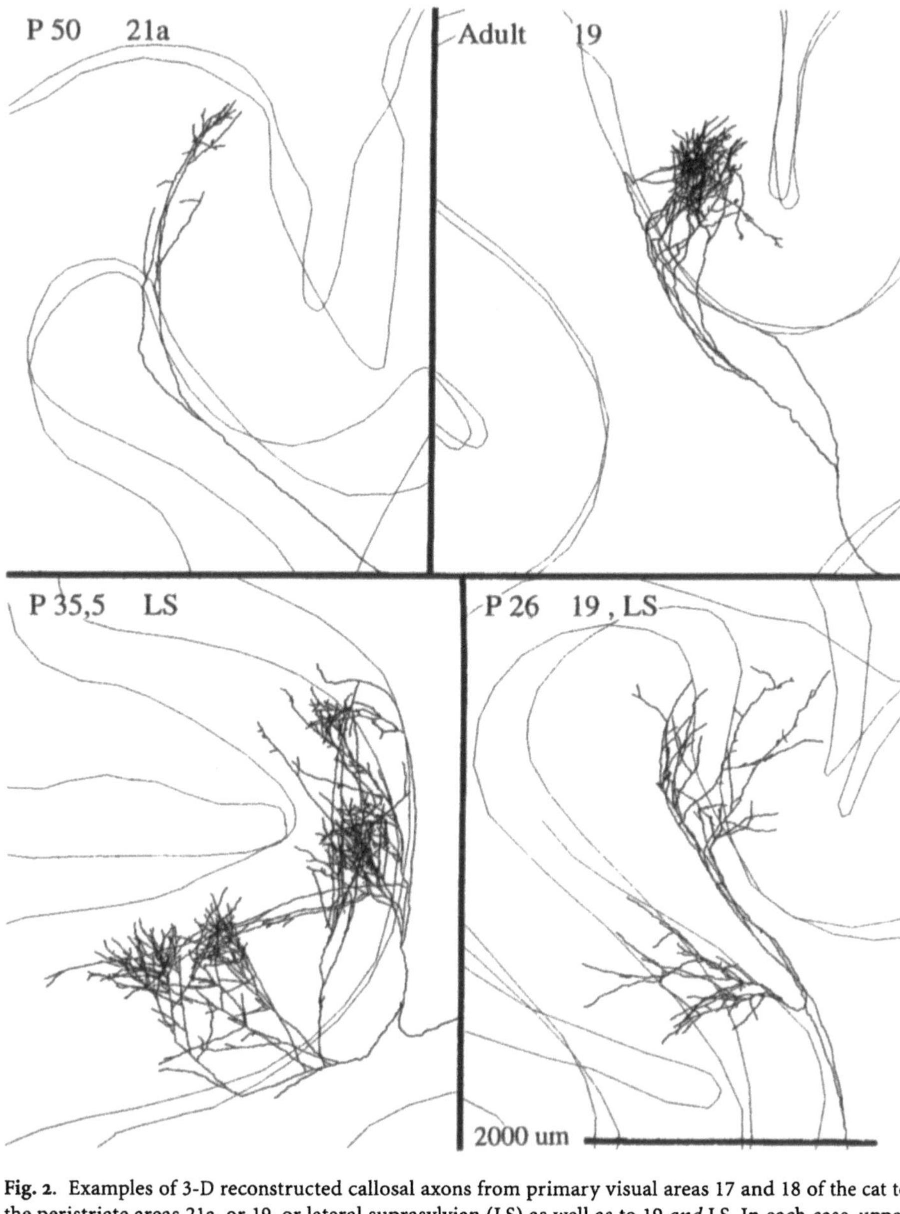

Fig. 2. Examples of 3-D reconstructed callosal axons from primary visual areas 17 and 18 of the cat to the peristriate areas 21a, or 19, or lateral suprasylvian (LS) as well as to 19 *and* LS. In each case, upper and lower boundaries of the gray matter in representative sections are also shown. The age of the animal is indicated in each case. (From Innocenti and Bressoud 1997)

nular layers. This selective laminar distribution of terminals suggests that callosal axons may preferentially target a subclass of cortical efferent neurons, presumably those that establish cortico-cortical connections.

It should be noticed that mapping is only one of the operations performed by

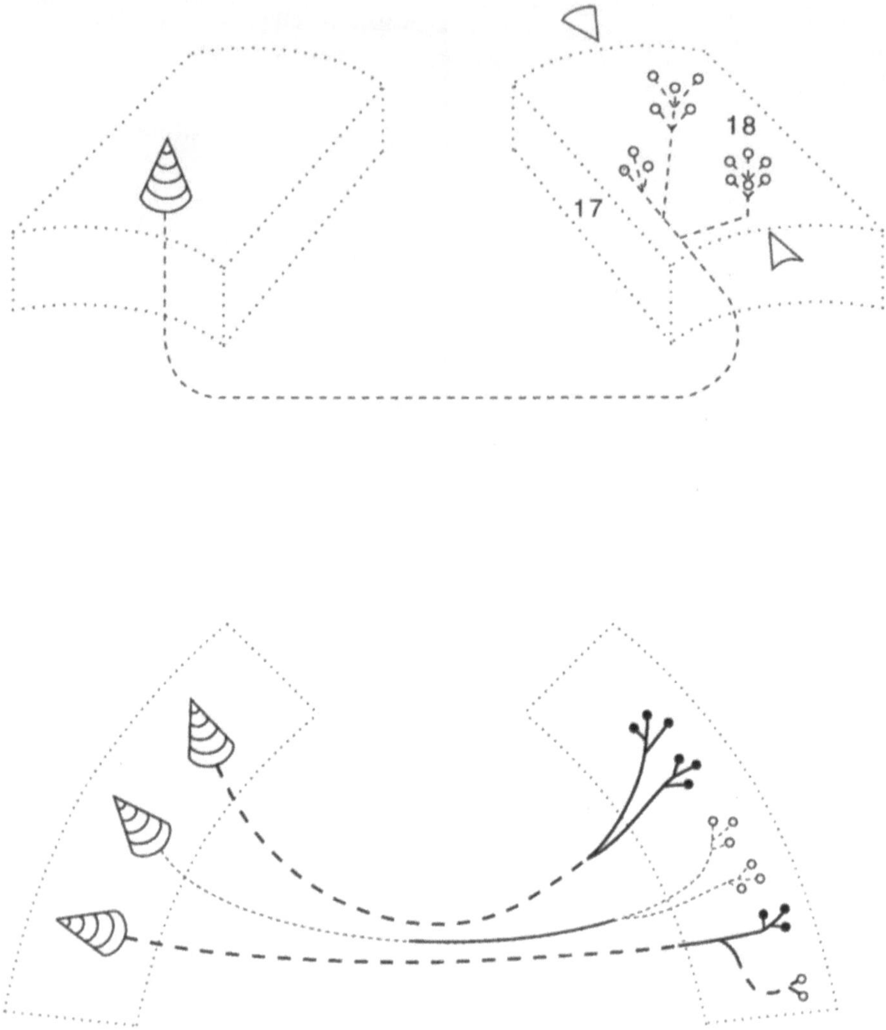

a

b

Fig. 3. Operations performed by callosal axons that interconnect areas 17 and 18 of the cat, by virtue of their geometry. **A** illustrates mapping and differential weighting. A point in one hemisphere, corresponding to the location of the cell body, is mapped into clusters of boutons (terminal columns) that are distributed in the contralateral cortex along the border between areas 17 and 18 (open triangles). Notice that the terminal columns contain different numbers of boutons, suggesting that the axon does not drive all the terminal columns with equal strength. **B** illustrates temporal transformations performed by axonal geometry. The top axon activates terminal columns simultaneously; the bottom axon activates terminal columns with a delay due to axonal conduction within its terminal arbor. The activation of the intermediate axon is retarded compared to the others, due to its smaller diameter. The active part of the axon is in solid black, the inactive part is interrupted. (From Innocenti and Bressoud 1997)

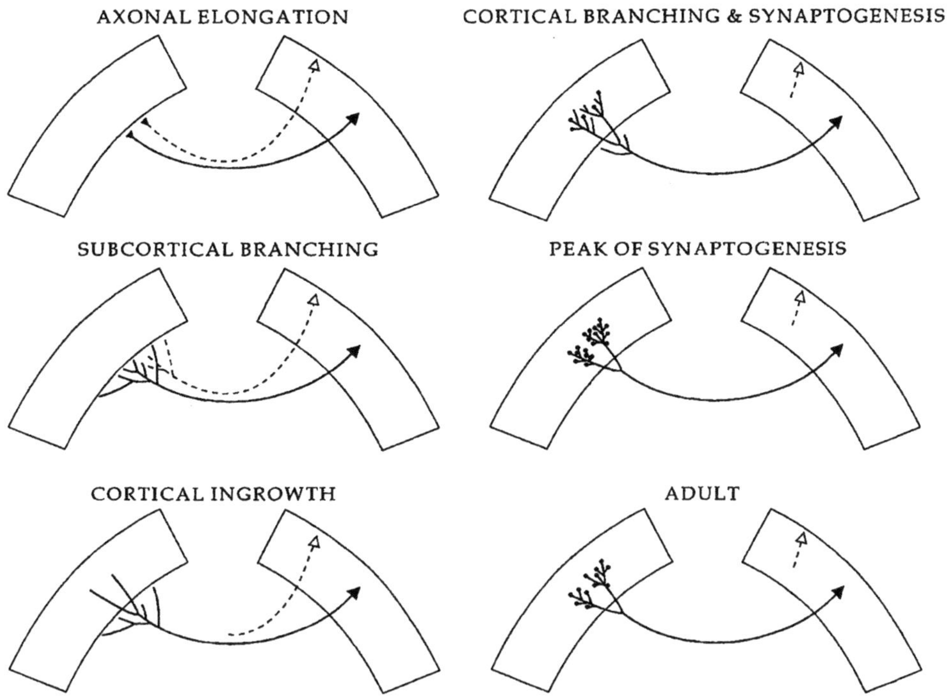

Fig. 4. Schematic representation of stages in the differentiation of callosal terminal arbors in visual areas of the cat. The stages are: axonal elongation, subcortical branching, growth into the gray matter, branching in the gray matter and synaptogenesis (here represented together). Notice that each stage is followed by partially regressive events, i. e., elimination of long callosal axons (interrupted) between the second and third stage, of subcortical branches during synaptogenesis and of synaptic boutons between the peak of synaptogenesis and adulthood. Neuronal cell bodies are represented by filled or open triangles, growth cones by small triangles at the end of the axonal branches, synaptic boutons by dots. Some of the neurons (open triangles) eliminate their callosal axons without dying, and establish permanent connections in the ipsilateral hemisphere, relatively close to their cell bodies (not shown). (From Innocenti and Bressoud 1997, modified)

an axon by virtue of its geometry. At least two others appear likely. One is the differential "weighting" of the spike transmitted along the axon, at the different terminal columns. This is due to the fact that the number of boutons in the columns can differ up to 50-fold for the same axon (Fig. 3). Therefore, probably different amounts of neurotransmitter are released at the different sites of axonal termination.

Another operation implemented by the axonal phenotype relates to the timing of neurotransmitter delivery (Fig. 3). Individual axons differ in their diameter and therefore in conduction velocity. Consequently, two spikes initiated simultaneously in different axons reach the targets asynchronously. In principle, the time of conduction to the different terminal columns of the same axon could also differ. Such delay-lines are an important property of axonal projections to the auditory brainstem (Carr and Konishi 1990). However, axons of this kind are rare among the callosal ones. The majority seems to adopt structural adjustments that achieve synchronous activation of their targets. These adjustments include diameter compensation for conduction length and exchange of branches among terminal columns (Innocenti et al. 1994).

In conclusion, callosal axons seem to implement mapping rules at several levels of cortical organization. Furthermore, the mapping takes place in a multi-dimensional space defined by neuronal properties other than location within the cortical volume. The connectional mapping implemented by individual callosal axons is the direct expression of their phenotypes. Axonal phenotype, however, implements more than connectional mapping. These multiple aspects of axonal phenotype are certainly not unique to callosal axons: they must be taken into account when dealing with the development of cortical connectivity.

Stages of Axonal Differentiation in the Development of Connections

As mentioned above, theories of the development of connections have traditionally stressed the role of cues guiding axons to their targets. This is undoubtedly an important aspect of the development of connections. However, given the multiple aspects of axonal phenotype and their functional implications, a more complete understanding of the development of connections requires a broader view on the differentiation of the axonal arbor. For visual callosal axons (Aggoun-Zouaoui and Innocenti 1994; Aggoun-Zouaoui et al. 1996), five stages of differentiation (Fig. 4) can tentatively be identified (reviewed in Innocenti 1995; Innocenti and Bressoud 1997). Briefly, **the first stage** consists in the elongation of axons between the hemispheres and up to the target cortex. **The second stage is** characterized by axonal collateralization in the white matter, underneath the target cortex. It might correspond to the so called "waiting period" described for thalamocortical axons (discussed in Aggoun-Zouaoui and Innocenti 1994). **The third stage** is characterized by the growth of callosal axons into the cortical gray matter and **the fourth stage** by the arborization in the cortical gray matter. **The fifth stage** is characterized by synaptogenesis.

The stages summarized above, are based almost exclusively on morphological criteria. Eventually, they will have to be validated by the identification of a more complete set of properties, including dynamic parameters of axonal growth as well as changes in the structural and molecular organization of the axon and its environment. They also do not cover the whole process of axonal differentiation. Well into the stage of synaptogenesis, the axons remain labile and can be either deleted or maintained depending on visual experience (see below). A molecular step might mark the transition between the juvenile/labile state and the adult/stable state. Elsewhere, we have speculated that the stabilization of the axons might require the maturation of cytoskeletal proteins including the heavy subunit of neurofilaments and the microtubule associated proteins Tau and MAP1b (Figlewicz et al. 1988; Guadaño-Ferraz et al. 1990). Myelination could represent a further stage of axonal differentiation, presumably a critical one for the temporal aspects of neural computation.

The Differentiation of Callosal Axons Entails Progressively Constrained Exuberant Growth and Selection

The stages of axonal differentiation listed above are the consequence of growth modes leading to elongation, collateralization, arborization and synaptogenesis, as proposed by Schneider et al. (1987) and Bhide and Frost (1991) for retinofugal axons. Each growth mode causes overproduction of axonal components, i.e., of callosal axons, of their branches in the white and in the gray matter, and of synapses (Fig. 4). These processes are followed by selection of axons, axonal branches and synapses to be maintained, and by elimination of the others. The elimination is massive. As mentioned earlier, at least 70 % of the juvenile callosal axons are eliminated in both cat and monkey (Berbel and Innocenti 1988; LaMantia and Rakic 1990). In the primary visual areas, many of the transient callosal axons originate from layer III and IV neurons, in the medial part of area 17. In contrast, in the adult, callosal projections originate exclusively near the 17/18 border. In addition, callosal axons interconnecting the 17/18 border regions have established, on average, about 10 branching points in the white matter during the second postnatal week, but these are reduced to less than two in the adult. The number of branches in the gray matter was found to exceed 500 at the beginning of the third postnatal month and it was down to around 240 in the adult. The number of boutons reached about 2100 at the beginning of the third postnatal month and it was down to the average adult value of about 300 by the end of the fifth month. It should be noticed that these values are only indicative because the number of serially reconstructed axons is limited. An important aspect of the exuberant growth is that, at each of the differentiation stages, the growth of individual axons is progressively more constrained within territories that gradually approximate their adult distributions.

The concept of exuberant growth described above probably applies to other cortical axons. In addition to the studies with retrogradely transported tracers (reviewed in Innocenti 1991), evidence of exuberant growth was obtained with anterograde tracers for geniculocortical axons (Ghosh and Shatz 1992; Antonini and Stryker 1993), callosal axons to the somatosensory areas of the rat (Norris and Kalil 1992; Halloran and Kalil 1994) and intra-areal axons (Callaway and Katz 1990; Assal and Innocenti 1993). The overproduction of synapses is a well-documented phenomenon in the development of cerebral cortex (reviewed in Rakic et al. 1994). This concept was established on the basis of electron-microscopic counts of density of synaptic boutons in sample volumes of cortical gray matter. However, the implications of synaptic overproduction for functional connectivity remained unclear, because the origin of the supernumerary boutons was undetermined. The analysis of callosal axons suggests that exuberant synaptogenesis is not incompatible with a topographically specific distribution of the connections (Aggoun-Zouaoui et al. 1996, and below). The synaptic reduction, therefore, does not appear to sharpen the topography of the connections and therefore it may not significantly affect the mapping functions implemented by the individual callosal axons. Instead, it might finely regulate the strength of the connections, by affecting the "weights" of action potentials transmitted by a single axon to its postsynaptic targets. This view is compatible with the available electrophysiological evidence (discussed in Aggoun-Zouaoui et al. 1996). However, uncertainties remain concerning the functional consequences of the overproduction of boutons since crucial parameters of synaptic function are unknown and the targets of the supernumerary boutons of callosal axons are unknown as well.

It is usually believed that exuberant growth demonstrates the potential plasticity of neural connections in development. In this perspective, it would seem that plastic changes in neural connectivity can occur at different stages of axonal differentiation, with different consequences for the organization of neural circuits.

Specific and Exuberant Growth: The Two Sides of Axonal Differentiation

We propose that the stages listed above are essentially stages of conditional differentiation. The axon is probably committed early to a given connectional phenotype, to a given sequence of differentiation stages and to specific growth modes. However, environmental cues appear to be necessary in order to 1) initiate the differentiation stages, 2) regulate the growth at each stage, and 3) select axonal components for maintenance or elimination. The exuberant growth appears to be an intrinsic property of the axon. It maximizes the probability that the axon will attain the cues on which its further differentiation depends.

This conceptual paradigm prompts the following questions. First, are cortical neurons committed to different connectional phenotypes? Second, if they are, how do they become committed? Third, what is the temporal relation between commitment and its phenotypic expression? Fourth, what is the nature of the interactions between a committed axon and its environment? Fifth, is axonal phenotype fully determined by early commitment or, alternatively, are certain aspects of axonal phenotype under epigenetic control?

The answers to the questions above are not at hand, but fragments of the answers exist for some of the questions.

Some restriction of connectional fate precedes axonal elongation

There can be little doubt that cortical neurons become committed to different connectional phenotypes very early, probably at the time of their last division. The strongest support for this concept is based on the evidence that cortical neurons differ in their connectional phenotypes along a radial axis extending from pia to white matter. Along this axis, two variables are mapped: neuronal birthdate, with older neurons generally at deeper positions, and connectional phenotype, with deeper neurons forming mainly subcortical projections and superficial neurons forming mainly cortical projections. Birthdate rather than radial position is strictly associated with the determination of connectional phenotype. Indeed, in the reeler mutant mouse, radial neuronal position is altered by defective migration, but connectional phenotype remains adequate for a given neuronal birthdate (Caviness 1977). Furthermore, neurons led to an abnormal radial position by experimental manipulations maintain the phenotype appropriate to their birthdate (Jensen and Killackey 1984; Innocenti and Berbel 1991; McConnell 1988). The mechanisms relating connectional phenotype to neuronal birthdate are under study. They might involve the expression of specific genes (Frantz et al. 1994).

Different connectional phenotypes are also expressed by neurons within the same layer. In the cat, neurons in layer III appear to establish, from the beginning, exclusive projections to one or the other cortical area (Innocenti and Clarke 1984). Indeed, as of the first postnatal week, some callosal axons are aiming specifically at the striate or at one of the extrastriate visual areas. Others are aiming at more than one area but almost exclusively at extrastriate visual areas, where such axons are also found in the adult. These connectional phenotypes remain identifiable throughout the whole period of axonal differentiation (Innocenti and Bressoud 1997; Bressoud and Innocenti, unpublished). Similarly, in the rat, the distinction between cortically and subcortically projecting layer V neurons is established very early, prior to the formation of an axon (Koester and O'Leary 1993).

The data mentioned are consistent with an early commitment of callosally projecting neurons to one of several connectional phenotypes.

Nevertheless, the existence of transient callosal and other cortico-cortical or corticofugal projections indicates that the full connectional phenotype is expressed late and progressively. As mentioned above, many of the callosal axons directed to primary and non-primary visual areas are transient and initially indistinguishable from those that will remain. Only at the later stages of subcortical branching and gray matter ingrowth can callosal axons destined to elimination, such as those from area 17, and axons destined to be maintained, such as those from the 17/18 border, be distinguished on morphological criteria (below). Similarly, cortico-pontine axons and transient cortico-spinal axons (O'Leary and Terashima 1988) are initially indistinguishable. The same is true for retinofugal axons (Bhide and Frost 1991), which initially grow beyond their targets and later collateralize into them.

Two alternative explanations exist for this axonal behavior. The first is that axons become committed to their connectional phenotype in stages, corresponding more or less to the stages of differentiation. The second, and more likely possibility, is that there might be a delay between the axonal commitment and its phenotypic expression.

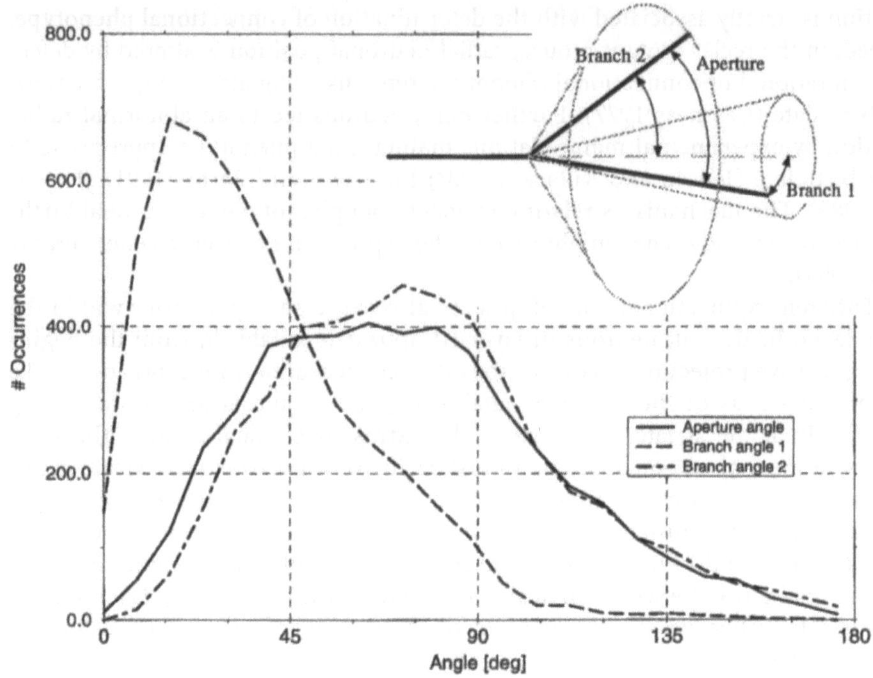

Fig. 5. Frequency of branching angles of 4650 bifurcation nodes of callosal axons ranging in age between postnatal day 3 and adult, and terminating in the gray matter near the border between visual areas 17 and 18. Branch 1 is by convention the branch with the smallest angle relative to the trajectory of the parent branch. Inset shows how the angles were measured (Tettoni et al., unpublished)

Intrinsic Growth Rules During Axonal Differentiation

The existence of growth rules intrinsic to the axon is suggested by the branching pattern of individual axons (Tettoni et al., unpublished). As shown in Figure 5, the angles of axonal branching are characteristic. One of the two daugther branches (branch 1) tends to follow the direction of the parent branch. The other (branch 2) deviates from the direction of the parent branch with an angle of about 70 % on average. Interestingly, this pattern of branching is found in all the axons we have examined thus far, irrespective of area of termination and of whether the branches are generated in the white or in the gray matter. This geometrical pattern, therefore, appears to reflect a growth rule that is likely to be intrinsic to the axon, possibly due to constraints imposed by the dynamics of cytoskeletal elements. This growth mode is compatible with a systematic "interstitial" branching (O'Leary and Terashima 1988). Presumably branch 1 is generated first and reflects the trajectory of elongation of the parent branch; branch 2 is generated later by interstitial budding.

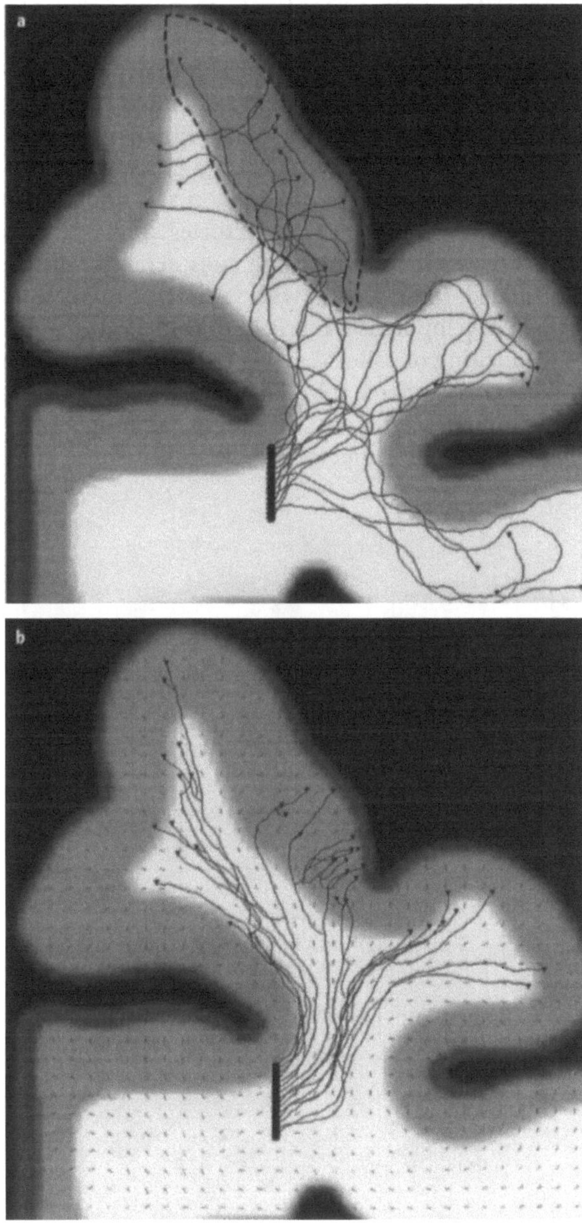

Fig. 6. Computer simulations of axonal growth on a bidimensional substrate with equal growth permissivity, restricted to the white matter and to the region of the gray matter within the interrupted contour (shown only in **A**). In **A**, axons grow randomly on an isotropic substrate, whereas in **B**, they align to cues (vector field) oriented roughly as radial glia. In **B**, branching was allowed. Small triangles at the end of the axons represent growth cones. (Tettoni and Innocenti, unpublished)

Axonal Differentiation Requires Cues in the Cellular Environment

The trajectories of callosal axons, at the stage of axonal elongation, appear to be influenced by cues in the substrate. This is suggested by the results obtained with computer simulations (Fig. 6) in which the axons advance at random on an evenly growth-permissive substrate modeled on the white matter under areas 17 and 18. The trajectories obtained in this situation are far less orderly that those observed in vivo. Instead, trajectories similar to the latter can be obtained when axons are guided by substrate cues arranged in curved lines resembling radial glia. Indeed, fasciculation with radial glia might guide callosal axons at this stage of their development (Norris and Kalil 1991).

Branching of callosal axons in the white matter also appears to be influenced by local cues. Indeed, it begins robustly only when axons have attained the white matter underlying areas 17 and 18, although some branches are formed elsewhere along their trajectory (Aggoun-Zouaoui and Innocenti 1994). The branch-

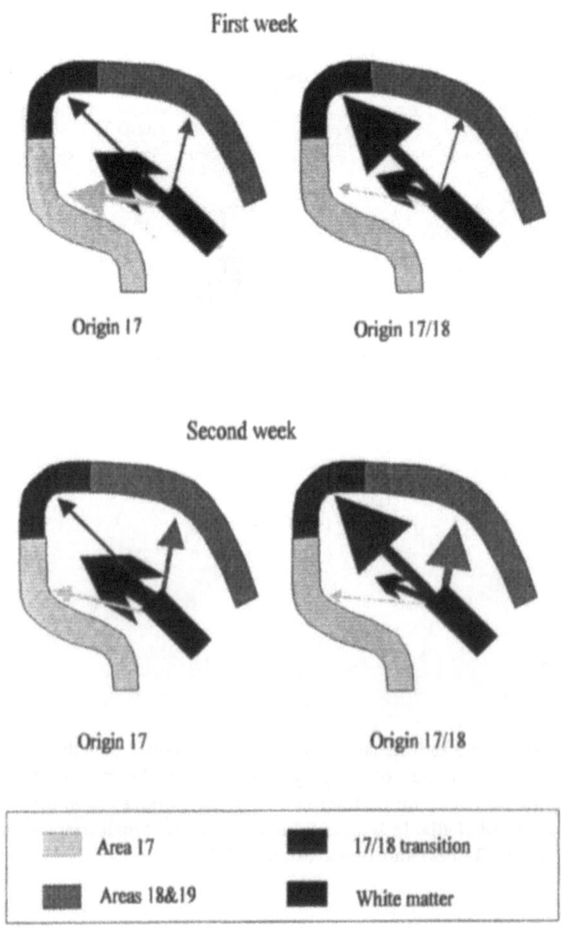

Fig. 7. Schematic representation of the distribution of branches of callosal axons with respect to the visual areas of the cat during the first and second postnatal weeks. Axons originating in area 17 are transient; those originating near the 17/18 border are all, or in the greatest majority, permanent. Thickness of arrows is proportional to the number of branches. (Based on Aggoun-Zouaoui and Innocenti 1994)

ing pattern varies depending on the origin and fate of the axons. The axons from area 17, which are destined to be eliminated, branch less abundantly and their branches terminate more often in the white matter without aiming clearly at one or the other cortical area. In comparison, axons from near the 17/18 border, most or all of which are destined to be maintained, branch more abundantly. The branches are more often aimed at the border between areas 17 and 18, rather than to area 17 (Fig. 7).

The stage of subcortical branching appears to precede the growth of axons into the cortex and might be a necessary condition for the latter to occur. Halloran and Kalil (1994) reported subcortical branching of callosal axons before they grow into the gray matter. We have also observed axonal branching in the white matter in the absence of branches growing into the gray matter but never the converse, i.e., axons in the gray matter that do not possess branches in the white matter (Aggoun-Zouaoui and Innocenti 1994).

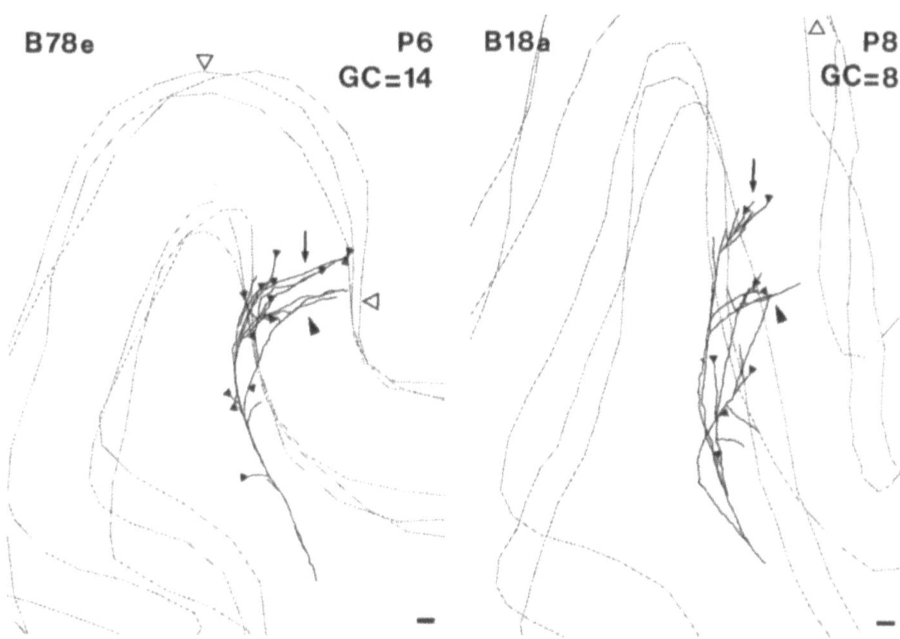

Fig. 8. Axonal arbors at the time of growth into the gray matter. Computer-aided 3-D reconstructions of axons originating in areas 17 and 18, labeled by biocytin and terminating near the border between areas 17 and 18. Contours of a few of the sections used for the reconstructions are also shown. Axons are coded as in Figure 1. Age of the animal at perfusion and total number of growth cones (GC) are indicated in the right-hand corner. In the reconstructions, growth cones appear as small triangles. Dorsal is up, medial to the left. Calibration bars are 100 μm. Arrows and arrowheads denote loose bundles of axons entering the gray matter in a columnar pattern. The transition between areas 17 and 18 is denoted by open triangles. (From Aggoun-Zouaoui et al. 1996, modified)

Local cues appear to regulate access of callosal axons to the gray matter, a process that begins around the end of the first postnatal week in the cat visual cortex. Only the region near the 17/18 border (and at some anteroposterior levels the whole of area 18, mediolaterally) allows ingrowth of callosal axons (Fig. 8). Only axons originating near the 17/18 border enter the gray matter. The vast majority of axons from area 17 remains confined to the white matter or to the subplate, although a few engage into the lower third of the gray matter in area 17, in 18 or near their common border. These axons, and the branches of axons from the 17/18 border directed to area 17, regress and disappear during the following weeks. This suggests that entering the gray matter might be a precondition for the further differentiation of the axons.

The importance of local cues at the stage of cortical ingrowth is further stressed by the fact that some axons enter the cortex with a disjunctive distribution of branches recalling the columnar pattern seen in the adult (Fig. 8). Probably, adult callosal axons interconnect columns detecting the same stimulus orientation (discussed by Houzel et al. 1994). Therefore, the early columnar ingrowth of callosal axons suggests that the axons recognize cues in the gray matter related to the early formation of cortical columns. Consequently, it provides tentative evidence that cortical columns begin to form before the onset of visual experience, and more than one week before they can be demonstrated with other means. The fact that thalamocortical axons are still rudimentary, at best, suggests that orientation columns might have a cellular substrate other than that imposed by afferent connections (discussed in Aggoun-Zouaoui et al. 1996).

Branching in the cortical gray matter begins almost as soon as the axons have entered it, i.e., at the end of the first postnatal week. In the following weeks, and until the beginning of the third postnatal month, the number of branches increases above that found in the adult. This exuberant growth blurs, somewhat, the previous disjunctive and columnar pattern of axonal ingrowth, but remains confined to the regions where callosal axons are found in the adult. In particular, branches are more numerous in certain layers, notably the supragranular layers. This resembles the layer-specific branching of corticofugal and of thalamocortical axons (Lund et al. 1977; Katz 1991; Bolz et al. 1992) and it might be due to intrinsic differences among the layers.

The differentiation of callosal axons near the 17/18 border might involve at least two kinds of local cues. One kind induces branching of callosal axons. The branching is specific for white matter regions underlying the target areas, although occasional branches can be found elsewhere along the trajectories of the axon. Branching of callosal axons might be induced by neurotrophic factors released near the site of axonal termination. Trophic factors appear to be involved in the development of the geniculo-cortical axons (Carmignoto et al. 1993; Cabelli et al. 1995; Galuske et al. 1996).

Other cues presumably regulate access of callosal axons into the gray matter. Growth-inhibiting signals (reviewed by Schwab et al. 1993) might prevent access of callosal axons to area 17 (discussed in Aggoun-Zouaoui and Innocenti 1994).

Activity Might Have a "Constructive" Role in Axonal Differentiation

Boutons, presumably synaptic boutons, appear in the gray matter at the end of the second postnatal week (Fig. 9). Over the following weeks, the number of boutons increases and, around the beginning of the third postnatal month, reaches a peak well above the adult values. The number of boutons decreases after the third postnatal month. The number of boutons of callosal axons evolves in parallel with the curves of overall synaptogenesis and synaptic reduction in cerebral cortex, as seen with electron-microscopic analysis (reviewed in Innocenti 1995).

Boutons acquire, from the beginning, a specific laminar distribution that is maintained throughout synaptogenesis. They appear initially in the infragranular layers, but then become particularly numerous in the supragranular layers. As is mostly the case in the adult, they are sparse or absent in layer IV. Boutons also acquire from the beginning a disjunctive columnar distribution that they maintain throughout synaptogenesis.

The specific distribution of synaptic boutons in the different cortical layers might be due to the differential molecular composition of the layers. This in turn might relate to the positional and connectional phenotypes of cortical neurons discussed above.

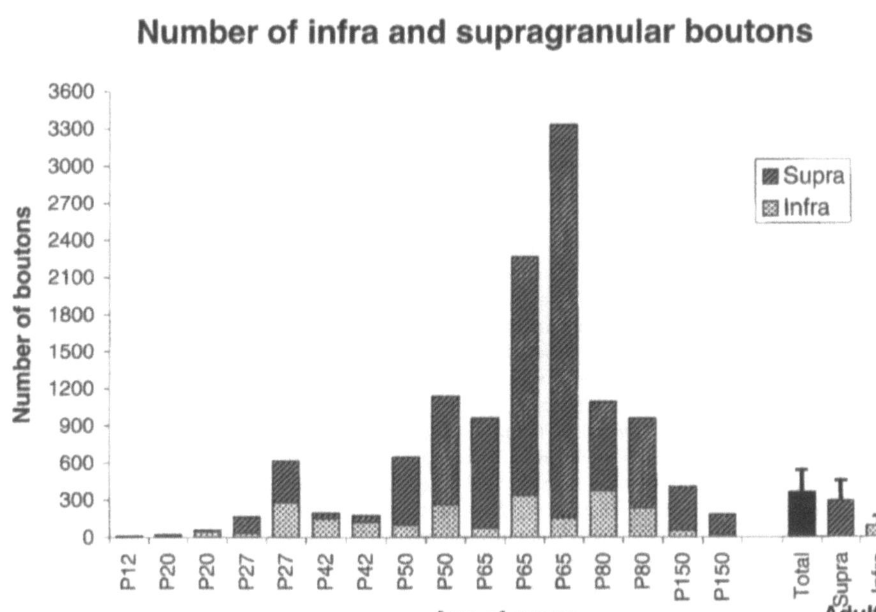

Fig. 9. Histogram showing the total number of boutons per axon in supragranular (supra) and infragranular (infra) sectors of the gray matter in axons reconstructed at different ages. Equal fractions of layer IV boutons are attributed to the two compartments. (From Aggoun-Zouaoui et al. 1996, modified)

The role of visual experience in maintaining callosal connections between the visual areas has been known for a while and was generalized to other cortico-cortical connections (reviewed in Innocenti 1991). Normal visual experience is necessary for maintaining the normal complement of callosal connections. In previous studies it was found that rearing kittens with bilateral eyelid suture resulted in the irreversible loss of about 50 % of the callosally projecting axons from areas 17 and 18 (Innocenti et al. 1985).

This finding suggested that the juvenile axons, axonal branches and synapses are labile structures. Visual experience appears to be involved in the selection and the stabilization of these labile structures.

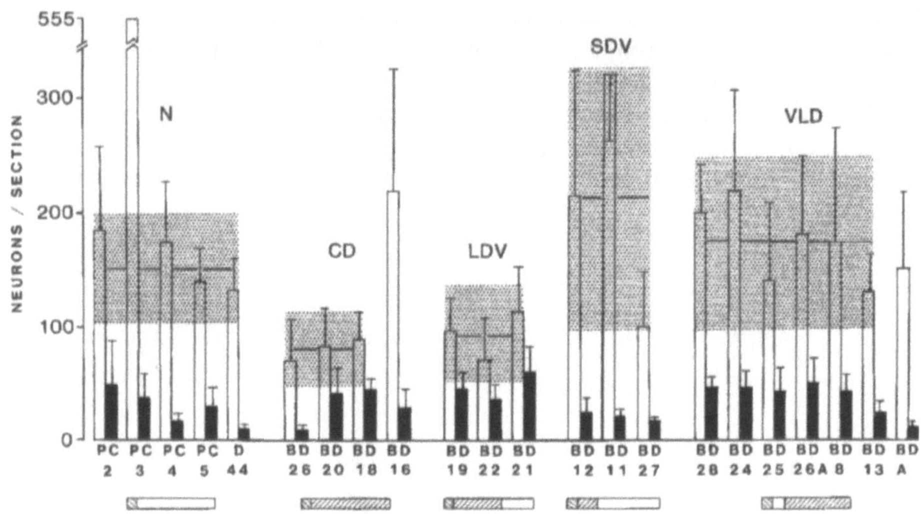

Fig. 10. Histograms showing the number of retrogradely labeled callosally projecting neurons in areas 17 and 18 of cats binocularly deprived of vision for different periods and at different ages. Each animal is denoted by code letters (PC, D, BD, A) and a number. Rearing conditions are indicated by the horizontal bars under the histograms. Hatching slanting downward from left to right indicates intervals during which the eyes were naturally closed. Hatching slanting downward from right to left indicates the period during which the eyelids were sutured shut. Open spaces indicate intervals when both eyes were open. N were normal cats. CD (continuous deprivation group) were cats continuously deprived of vision for the duration of their lives (BD 16 until postnatal day 42). LVD (long deprivation followed by vision group) were cats deprived of vision until day 90; their eyes were then reopened and they were allowed normal vision for about two months. SDV (short deprivation followed by vision group) were cats deprived of vision until postnatal day 37 or 38; their eyes were then reopened and they were allowed normal vision until the end of the fifth month. VLD (vision followed by long deprivation group) had normal vision until postnatal days 17 to 31, following natural eye opening at the end of the first week. BDA was and adult cat deprived of vision for three months. The vertical axis indicates the number of labeled neurons per section. Open vertical bars represent counts for neurons in layers II–IV; solid vertical bars represent counts for layers V and VI. Vertical lines represent one standard deviation (SD) in the counts for individual cats. Horizontal lines between the open vertical columns indicate the mean number of labeled callosal neurons per section for the groups, while stippling represents the region of 1 SD from the mean. BDA and BD16 were excluded from these statistics. (From Innocenti et al. 1985, with permission)

It is interesting to compare the sensitive period of callosal axons to visually driven activity with the curves of bouton production by callosal axons and of cortical synaptogenesis. To identify when callosal connections can be stabilized by visual experience, kittens were raised with their eyes bilaterally sutured shut for different periods of time before or after normal eye opening (Innocenti et al. 1985; Fig. 10). In one of the experimental series, kittens were allowed short periods of normal vision before their eyelids were sutured (VLD, Fig. 10). It was found that 10 days of normal vision following natural eye opening at Postnatal day (P) 7–10 were sufficient to prevent the loss of callosal axons in subsequent periods of binocular deprivation, no matter how long the periods were. It is puzzling that in the 10 days following eye opening, callosal axons appear to have only begun to establish the first synapses. This protective role of vision can be delayed, since kittens deprived of vision until the beginning of the first postnatal months and then allowed normal vision (SDV; Fig. 10) also did not lose callosal connections.

These observations suggest that vision begins to affect the development of callosal connections during the period of increasing synaptic number rather than, as expected, during the phase of reduction. Probably, visual activity can validate/eliminate synapses soon after they have formed. It might even stimulate the formation of new synapses. This implies that activity might have a "constructive" role in the development of connections. That is, axonal arbors might be finely shaped by ongoing activity, while they are being constructed. Unfortunately, it is not known at what age the earliest consequences of visual deprivation can be detected and how axonal arbors differentiate in the absence of normal visual activity.

Conclusions

The morphological and molecular analyses provide the necessary background information about developmental processes but are insufficient to capture their complex dynamics. In particular, the development of connections seems to involve complex and delicately balanced interactions among several types of molecules, favoring or inhibiting axonal growth (Tessier-Lavigne and Goodman 1996). Times might be ripe for understanding the development of neural connections at the algorithmic level. This level of analysis requires quantitative investigations of the operations performed by the growing axon in its cellular environment. Subsequently, these operations can be addressed in numerical simulations of axonal growth. This approach might capture the dynamics of the development of connections and, in addition, could have predictive power on the underlying morphological and molecular events.

Our analysis has focused on the overall structure of the processes leading to the development of one type of cortical connections as well as on the nature of the interactions between differentiating axons and their environment. We suggested that the axons execute growth/differentiation "programs." These programs appear to be of an interactive kind requiring selective cues in the environment, although this does not exclude the possibility that instructive cues might be used as well. The difference between selective and instructive cues can be illustrated by a metaphor derived from two popular children's games. In one of the games, "treasure hunting," the cues provide detailed instructions about the direction of search. In the other, "hide-and-seek," the cues are rather yes/no instructions that eventually canalize the search by providing information about the error generated by groping around in different directions. For some reason, the latter is the game that evolution has decided growing axons ought to play. This might not be accidental. One of us argued elsewhere (Innocenti 1995) that axonal differentiation by exuberant growth regression could have had a permissive role in the evolution of cerebral cortex. Therefore they were widely preserved across phyla.

The "hide-and-seek" metaphor implies that the yes/no instructions promoting or inhibiting axonal growth are delivered sequentially. The sequence of instructions could be determined by: 1) spatial separation of cues-containing signposts, for example cell groups that produce growth enhancing or inhibiting signals; 2) sequence of maturation of cues in the cellular environment, for example the maturation of the cell groups generating the above-mentioned signals; 3) sequence of axonal maturation, with certain signals not being recognized until the axon has developed the appropriate receptors. Exuberant growth maximizes the probability that the axon will attain the cues on which its further differentiation depends by favoring the search of spatially distributed cues and allowing the axon to "wait" during the delays imposed by the maturational processes.

Interestingly though, in light of what was discussed above, a Darwinistic metaphor (Edelman 1987) seems inappropriate to describe the regressive events that characterize the interaction between growing cortical axons and their environment. Darwin's natural selection acted on fully differentiated phenotypes, varied by genetic mutation. It constrained and decreased the diversity of phenotypes by eliminating the least fit ones. By contrast, developmental selection acts on differentiating nerve cells and circuits that it sculpts out by selective deletions of axons, axonal branches, etc. Therefore, it increases phenotypic variation. The cellular environment appears to have the power of canalizing axonal differentiation towards one phenotype within a possible set whose size remains to be determined.

Acknowledgments

Supported by Swiss National Science Foundation grant 31-39707.93 to GMI. We warmly thank Mrs M. Blanc for her help with this text, and E. Bernardi and Miss S. Massonet for their art work.

References

Aggoun-Zouaoui D, Innocenti GM (1994) Juvenile visual callosal axons in kittens display origin- and fate-related morpholgy and distribution of arbors. Eur J Neurosci 6:1846–1863

Aggoun-Zouaoui D, Kiper DC, Innocenti GM (1996) Growth of callosal terminal arbors in primary visual areas of the cat. Eur J Neurosci 8:1132–1148

Antonini A, Stryker MP (1993) Development of individual geniculocortical arbors in cat striate cortex and effects of binocular impulse blockade. J Neurosci 13:3549–3573

Assal F, Innocenti GM (1993) Transient intra-areal axons in developing cat visual cortex. Cereb Cortex 3:290–303

Berbel P, Innocenti GM (1988) The development of the corpus callosum in cats: a light- and electron-microscopic study. J Comp Neurol 276:132–156

Bhide PG, Frost DO (1991) Stages of growth of hamster retino fugal axons: implications for developing axonal pathways with multiple targets. J Neurosci 11:485–504

Bolz J, Novak N, Staiger V (1992) Formation of specific afferent connections in organotypic slice cultures form rat visual cortex cocultured with lateral geniculate nucleus. J Neurosci 12:3054–3070

Cabelli RJ, Hohn A, Shatz CJ (1995) Inhibition of ocular dominance column formation by infusion of NT-4/5 or BDNF. Science 267:1662–1666

Callaway EM, Katz LC (1990) Emergence and refinement of clustered horizontal connections in cat striate cortex. J Neurosci 10:1134–1153

Carmignoto G, Canella R, Candeo P, Comelli MC, Maffei I (1993) Effects of nerve growth factor on neuronal plasticity of the kitten striate cortex. J Physiol 464:343–360

Carr CE, Konishi M (1990) A circuit for detection of interaural time difference in the brain stem of the barn owl. J Neurosci 10:3227–3246

Caviness VS, Jr (1977) Reeler mutant mouse: a genetic experiment in developing mammalian cortex: In: Cowan WM, Ferrendelli JA (eds) Approaches to cell biology of neurons. Society for Neuroscience, Bethesda, MA, pp 27–46

Changeux JP, Danchin A (1976) Selective stabilisation of developing synapses as a mechanism for the specification of neuronal networks, Nature 264:705–712

Cramer KS, Sur M (1995) Activity-dependent remodeling of connections in the mammalian visual system. Curr Opinion Neurobiol 5:106–111

Easter Jr SS, Purves D, Rakic P, Spitzer NC (1985) The changing view of neural specificity. Science 230:507–511

Edelman GM (1987) Neural Darwinism. 1st Edition. Basic Books, Inc., New York

Figlewicz DA, Gremo F, Innocenti GM (1988) Differential expression of neurofilament subunits in the developing corpus callosum. Dev Brain Res 42:181–189

Frantz GD, Weimann JM, Levin ME, McConnell SK (1994) Otxl and Otx2 define layers and regions in developing cerebral cortex and cerebellum. J Neurosci 14:5725–5740

Galuske RAW, Kim D-S, Castrén E, Thoenen H, Singer W (1996) Brain-derived neurotrophic factor reverses experience-dependent synaptic modifications in kitten visual cortex. Eur J Neurosci 8:1554–1559

Ghosh A, Shatz CJ (1992) Pathfinding and target selection by developing geniculocortical axons. J Neurosci 12:39–55

Goodman CS, Shatz CJ (1993) Developmental mechanisms that generate precise patterns of neuronal connectivity. Cell 72:77–98

Guadano-Ferraz A, Riederer BM, Innocenti GM (1990) Developmental changes in the heavy subunit of neurofilaments in the corpus callosum of the cat. Dev Brain Res 56:244–256

Halloran MC, Kalil K (1994) Dynamic behaviors of growth cones extending in the corpus callosum of living cortical brain slices observed with video microscopy. J Neurosci 14:2161–2177

Houzel J-C, Milleret C, Innocenti G (1994) Morphology of callosal axons interconnecting areas 17 and 18 of the cat. Eur J Neurosci 6:898–917

Innocenti GM (1986) General organization of callosal connections in the cerebral cortex. Cereb Cortex 5:291–353

Innocenti GM (1988) Loss of axonal projections in the development of the mammalian brain In: Parnavelas JG, Stern CD, Stirling RV (eds) The making of the nervous system. Oxford University Press, Oxford, pp 319–339
Innocenti GM (1991) The development of projections from cerebral cortex. In: Progress in sensory physiology. Vol. 12. Springer-Verlag, Berlin, pp 65–114
Innocenti GM, Lehmann P, Houzel J-C (1994) Computational structure of visual callosal axons. Eur J Neurosci 6:918–935
Innocenti GM (1995) Exuberant development of connections, and its possible permissive role in-cortical evolution. Trends Neurosci 18:397–402
Innocenti GM, Berbel P (1991) Analysis of an experimental cortical network: ii) Connections of visual areas 17 and 18 after neonatal injections of ibotenic acid. J Neural Transplant Plast 2:29–54
Innocenti GM, Bressoud R (1997) Callosal axons as computational devices. In: Zaidel E, Iacoboni M, Pascual-Leone A (eds) Structure and function of the human corpus callosum: a case study incognitive neuroscience. Plenum Press, New York, in press
Innocenti GM, Clarke S (1984) The organization of immature callosal connections. J Comp Neurol 230:287–309
Innocenti GM, Frost DO, Illes J (1985) Maturation of visual callosal connections in visually deprived kittens: a challenging critical period. J Neurosci 5:255–267
Jensen KR, Killackey HP (1984) Subcortical projections from ectopic neocortical neurons. Proc Natl Acad Sci USA 81:964–968
Katz LC (1991) Specificity in the development of vertical connections in cat striate cortex. Eur J Neurosci 3:1–9
Koester SE, O'Leary DDM (1993) Connectional distinction between callosal and subcortical projecting cortical neurons is determined prior to axon extension. Dev Biol 160:1–14
LaMantia A-S, Rakic P (1990) Axon overproduction and elimination in the corpus callosum of the developing rhesus monkey. J Neurosci 10:2156–2175
Lund JS, Boothe RG, Lund RG (1977) Development of neurons in the visual cortex (area 17) of the monkey (*Macaca nemestrina*): a Golgi study form fetal day 127 to postnatal maturity. J Comp Neurol 176:149–187
McConnell SK (1988) Fates of visual cortical neurons in the ferret after isochronic and heterochronic transplantation. J neurosci 8:945–974
Norris CR, Kalil K (1991) Guidance of callosal axons by radial glia in the developing cerebral cortex. J Neurosci 11:3481–3492
Norris CR, Kalil K (1992) Development of callosal connections in the sensorimotor cortex of the hamster. J Comp Neurol 326:121–132
O'Leary DDM (1992) Development of connectional diversity and specificity in the mammalian brain by the pruning of collateral projections. Curr Biol 2:70–77
O'Leary DDM, Terashima T (1988) Cortical axons branch to multiple subcortical targets by interstitial axon budding: implications for target recognition and "waiting periods." Neuron 1:901–910
Rakic P, Bourgeois J-P, Goldman-Rakic PS (1994) Synaptic development of the cerebral cortex: implications for learning, memory, and mental illness. Prog Brain Res 102:227–243
Rétaux S, Harris WA (1996) Engrailed and retinotectal topography. Trends Neurosci 19:542–546
Schneider GE, Jhaveri S, Davis WF (1987) On the development of neuronal arbors. In: Chagas C, Linden R (eds) Working group on developmental neurobiology of mammals. Pontificia Academia Scientiarum, Citta del Vaticano
Schwab ME, Kapshammer JP, Bandtlow CE (1993) Inhibitors of neurite growth. Ann Rev Neurosci 16:565–595
Segraves MA, Rosenquist AC (1982) The afferent and efferent callosal connections of retinotopically defined areas in cat cortex. J Neurosci 2:1090–1107
Singer W (1995) Development and plasticity of cortical processing architectures. Science 270:758–764
Sperry RW (1963) Chemoaffinity in the orderly growth of nerve fiber patterns and connections. Proc Natl Acad Sci USA 50:703–710
Tessier-Lavigne M, Goodman CS (1996) The molecular biology of axon guidance. Science 274:1123–1133

What about Isocortex can be Rewired and Reconfigured?

J.K. Niederer, M.A. Kingsbury and B.L. Finlay

Abstract

The mammalian isocortex is an intriguing mix of structural consistency and functional diversity. Across mammals, isocortex configuration is conserved to a very high degree, with visual cortex posterior, somatosensory cortex medial, and the associated secondary areas proximal to these primary areas. The classic, six-layered internal cortical architecture is conserved. Yet, cortical areas are also structurally and functionally distinct and can be distinguished on the basis of architecture, biochemistry, physiology and connectivity. Across species, the differences between areas are remarkably consistent. Is there something special about each cortical area that makes it adept at handling its particular input, or is it just a piece of a calculating machine that is indifferent to the modality of its information and simply performs routine transformations? If there are true differences in the cortical mosaic, when in development do they arise, and what feature of gene expression, connectivity or function carries the information that diversifies cortical areas?

The Ventricular Zone, Subplate, and First Fiber Contacts

The general features of cortical development and specification have been much reviewed (Götz 1995; Kennedy and Dehay 1993; O'Leary et al. 1994). We recount the basic facts here to set context and highlight those features that we will concentrate on in experimental perturbations of cortical development. Neurons that form the six layers of isocortex are derived from the dorsal aspect of the lateral ventricles. The first-generated cells form the preplate (Marin-Padilla 1971), which is split into two layers, cortical layer 1 and the transient subplate (Luskin and Shatz 1985), by successive waves of generated cells. New neurons migrate over the most recently settled cells to take up their place underneath layer 1 and make up the cortical layers 2 through 6 (Angevine and Sidman 1961). Though the rules of laminar development appear straightforward (migrate up to the bottom of layer 1 and settle), laminar fate seems to be programmed within the ventricular zone. McConnell and her colleagues have shown this by removing replicating cells from the ventricular zone of one animal and transplanting them into the neuroepithelium of another animal at a different developmental stage (McCon-

A. M. Galaburda / Y. Christen (Eds.)
Normal and Abnormal Development of Cortex
© Springer-Verlag Berlin Heidelberg New York 1997

nell 1985, 1988). Transplanted cells that migrate end up in the deeper layers appropriate to their site of origin, rather than the migratory position appropriate to their new context. Since different cortical layers have different cell types in the mature cortex, McConnell argues that, either directly or indirectly, cell type can be determined before cortical cells leave the ventricular zone (McConnell 1989).

The first generated cells of the developing cortex, the transient subplate layer, have been shown to be important in guiding or permitting axons from the thalamus to find their targets. Subplate neurons are the first mature neurons produced by the neuroepithelium and begin to sprout projections almost immediately after they are born. They project into the internal capsule, where they encounter thalamic axons on their way to the cortex. The nature of this encounter is the subject of some debate (Miller et al. 1993; Molnár and Blakemore 1995). Both thalamocortical and corticothalamic projections show a very high maintenance of topographic order in their projections independent of each other, and the central question is how the two are brought in register. Thalamic axons fasciculate at least in part with corticofugal projections within the white matter, which could provide a convenient method for guidance of the thalamocortical axons to their target region (Ghosh and Shatz 1992). When subplate cells are killed by the excitotoxin kainic acid before thalamic axons reach their targets (Ghosh et al. 1990; Ghosh and Shatz 1993), thalamic axons grow past their appropriate cortical area. However, studies from our own laboratory and other laboratories show two separate pathways, with different trajectories, that share an apposed face at some point during their development (Miller et al. 1993; De Carlos and O'Leary 1992). Other researchers have also found that the pathways are indeed clearly separated in both developing systems (Bicknese et al. 1994; De Carlos and O'Leary 1992) and in the adult (Bicknese et al. 1994; De Carlos and O'Leary 1992; Woodward et al. 1990). This finding has recently been supported by coculture studies showing that thalamocortical and corticothalamic axons actually repel each other (Bagnard et al. 1995). Information that can resolve this debate involves whether there are perhaps two separate groups of confuseable axons, subplate and layer 5, that in fact have different behaviors. If subplate axons are not guiding the axons to their target, perhaps they are marking the thalamic axons' appropriate address (Ghosh and Shatz 1992). It is reasonable to search for sources of information other than subplate that might organize both the thalamocortical and corticothalamic axons into topographic correspondence.

One of the first events that occurs during the development of the cortex is the production of radial glia fibers. These fibers connect the bottom of the ventricular zone to the top edge of the brain, the pia, and establish a scaffold upon which the immature cortex can be built. These radial glia fibers are in an ideal position to guide newly generated neurons into their proper location in the cortical plate, as Rakic has suggested (Rakic 1978). Neurons produced in the ventricular zone migrate up their local glial guide to their appropriate position in the cortical plate. Newly produced neurons are found in close apposition to radial glial cells (Levitt and Rakic 1980; Misson et al. 1991; Rakic 1971, 1972). If positional information is already in place in the ventricular zone, one or a small number of proliferative units in the ventricular zone constrained in their migration could determine the positional marker of an entire cortical cell column, the "proto-map" theory of Rakic (1988). This hypothesis is supported by the differential staining of immunocytochemical markers in the early ventricular zone (Levitt et al. 1981, 1983; Levitt and Rakic 1980). Areal specificity could be determined relatively simply within the ventricular zone and without subcortical input.

Much evidence suggests that the protomap hypothesis would require substantial qualification. In culture, using confocal microscopy, newly produced neurons in the ventricular zone are shown to migrate tangentially for impressive distances before settling into a particular cortical column (O'Rourke et al. 1992). In vivo studies by Cepko and Walsh (Walsh and Cepko 1990, 1992, 1993) have shown that retrovirally labelled neurons in the developing cortex have the capability to disperse tangentially. Clonally related cells are capable of dispersing wide distances in the cortex, well past areal boundaries (Walsh and Cepko 1992), arguing against simple radial migration as the singular mode of cortical develop ment. Other retrovirus experiments have found similar results (Luskin et al. 1988; Price and Thurlow 1988; Walsh and Cepko 1992), although with less impressive dispersal distances [200–300 μm as opposed to up to several centimeters in Walsh and Cepko (1992)].

Experimental Demonstrations of Equipotentiality

If early isocortex is truly equipotential, it should be able to accept thalamic input of any modality and perform some kind of routine transformation. Sur and his colleagues have tested this hypothesis (Sur et al. 1988). By removing a cortical area's normal input, and taking away the normal target for a subcortical projection, they were able to force visual information into auditory cortex. After denervating the medial geniculate nucleus (MGN) of its auditory input early in development, visual input was induced to invade the MGN by indirectly removing the retinal projection's thalamic target, the dorsal lateral geniculate nucleus (dLGN). This was accomplished by removing the visual cortex, which leads to a degeneration of most of the dLGN. MGN then received part of the visual projection from the retina and made its normal projection to the primary auditory cortex. Through this procedure, the auditory cortex becomes the recipient of visual information and a rudimentary map of visual space is created in the novel cortex (Pallas and Sur 1993). These results indicate that isocortex may be a generic processing structure.

Another line of evidence showing the plasticity of isocortex comes from studies of the rodent somatosensory maps. The whisker representation, or barrelfield, in many rodent species takes up a large proportion of somatosensory cortex. Within a species, the maps of these barrels are remarkably consistent. Barrels are dependent upon their input for proper formation and are capable of a high level of developmental flexibility. If a row of whiskers is removed early in the development of an animal, other barrels are able to expand into the unoccupied cortical space that normally would have been designated to the missing whiskers (Woolsey and Wann 1976). Mice that are binocularly enucleated show an increase in the size of their barrelfields (Rauschecker et al. 1992). This finding shows that, within a given cortical area, subcortical input may be capable of driving the parcellation of cortical space.

Further evidence of isocortex equipotentiality comes from transplantation studies (Stanfield and O'Leary 1985). O'Leary and Stanfield have shown that embryonic occipital cortex removed from one animal and transplanted into medial cortex of a newborn can be sustained and take on many of the properties of its host area. The transplants were able to make a sustained projection into the spinal cord, a feature of medial cortex, but not of occipital cortex. When the experiment was repeated using transplants taken from newborn occipital cortex, in which subcortical input had already been established, the transplanted tissue was unable to make a sustained projection to the spinal cord. Schlaggar and O'Leary have done additional transplantation experiments (Schlaggar and O'Leary 1991) to assess whether presumptive visual cortex would be able to form barrels if placed in the barrelfield cortex of primary somatosensory cortex. They found that, within the transplanted visual cortex, typical barrel-like projections from the somatosensory thalamus (ventrobasal nucleus, VB) into the transplanted cortex were formed. Together, these experiments show that properties associated with a given cortical area are not functionally defined by where the cells of the area were produced. These experiments also implicate subcortical information as the determining factor in cortical differentiation. In a theory called the "Protocortex Model" O'Leary (1989) postulated from these results that cortical differentiation occurs only with the influence of subcortical projections, and the subcortical input that has been shown to be the most likely candidate is the thalamus.

We have been investigating the connectional specificity of cortex in the absence of thalamic input as a way of examining an "uninstructed cortex," rather than instructional mismatch. The results are striking and consistent. Kingsbury et al. (1995) have shown that when thalamic nuclei (either principally lateral geniculate or principally ventrobasal) are removed early in development, intracortical projections at maturity are more numerous, more widespread and change in their normal laminar distribution. An example of such an animal with an early lateral geniculate lesion is shown in Figure 1. Since cortical neurons project somewhat diffusely, later refining their projections to the mature pattern (Callaway and Katz 1990), it seems likely that these extended projections represent a stabilized initial state, though this remains to be conclusively demonstrated. The same is true for the extent and number of callosal projections after thalamic lesions (Miller et al. 1991). Most interesting of all, subcortical projections appear to be maintained in their initial state. For example, after early lateral geniculate lesions, a projection can be found at maturity from V1 past the caudal medulla, and the normal visual corticopontine projection loses its topographic specificity, covering a pontine volume more than ten times larger than its normal state (Kingsbury and Finlay, unpublished observations). Thus thalamic information appears to be crucial for the normal development and retraction of both

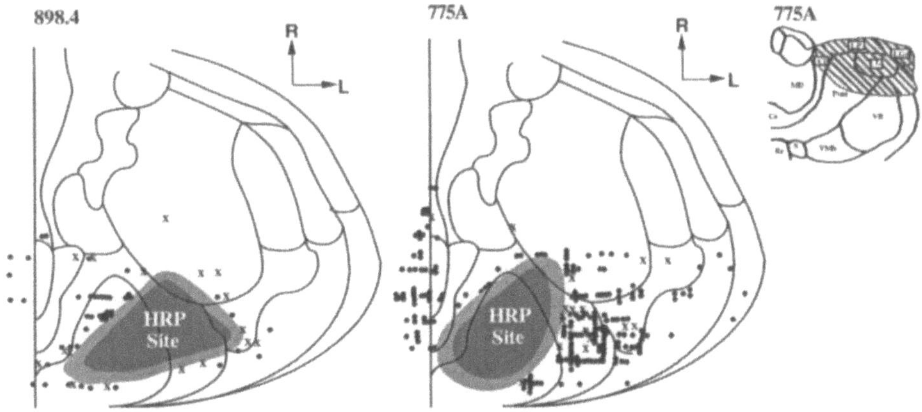

Fig. 1. Dorsal view reconstructions of adult hamster cortex showing the distribution of retrogradely labeled intracortical cells projecting to visual cortex in a control animal (left) and in an animal that received an ablation of visual thalamic nuclei on P0 (right). Circles and Xs represent cells labeled in the infragranular layers and supragranular layers, respectively. The inset is a coronal section that depicts the extent of thalamic damage in the lesioned animal. HR-horseradish peroxidase

intracortical and subcortical projections. Whether this information takes the form of molecular signals passed from thalamus to cortex or the patterning of thalamic activity is unknown.

Non-Equipotentiality in the Ventricular Zone

From very early in development, markers that delineate borders of general cortical regions have been shown to exist. One such marker is the limbic-associated membrane protein (LAMP), expressed early in corticogenesis, which differentiates limbic cortex from other cortical areas (Barbe and Levitt 1991, 1992). If limbic cortex is transplanted to another cortical area early in development, it will continue to express the marker and will also receive cortical afferents typical of its origin, rather than that of its host (Barbe and Levitt 1992). This finding demonstrates the possibility that thalamic afferents use local markers to find their correct targets, in the absence of a subplate guide. Of course, limbic cortex is a fundamentally different type of cortex than isocortex, with a different lamination pattern, and these results might be irrelevant to the problem of parcellation within isocortex. There are, however, clear molecular differences expressed entirely within the boundaries of isocortex (Cohen-Tannoudji et al. 1994). One marker delineates the boundaries of the future somatosensory cortex early in the developing embryo. Transplanting cortex into other regions of the brain does not affect the expression of this marker, and it is expressed before thalamocortical axons arrive in the cortex. Johnston and van der Kooy (1989) have found differential levels of protooncogene expression early in the developing isocortex. These studies show that there is clear neurochemical heterogeneity in the isocortex ventricular zone.

There are other local disparities in addition to gene and protein expression. Primary visual cortex, especially in the primate, has a markedly greater cell density than any other cortical area (Dehay et al. 1991). Post-migration and post-innervation factors, such as differential cell death among the various cortical areas, contribute to but do not entirely account for the differences in the numbers of neurons in a cortical column (Finlay 1992; Finlay and Slattery 1983). Kennedy and Dehay (Dehay et al. 1991) have shown that there are many more cells produced in the ventricular zone directly below presumptive primary visual cortex than underneath neighboring secondary visual cortex. Bayer and Altman (1991) have reported similar findings in the rodent motor cortex. When layer 5 cells are being produced in the ventricular zone, there are many more cells underneath motor cortex than there are under visual cortex, even allowing for neurogenetic gradients. More recent studies in rodents have extended these findings to primary and secondary somatosensory areas (Polleux et al. 1995), where differences in neuron production rates were found among three separate areas in the ventricular zone. Together, these results strongly suggest that each area in the cortex, or at least those associated with sensory processing, has already begun differentiation by mechanisms within the neuroepithelium, independent of external influences.

The Role of the Thalamus in Cortical Development

Does evidence showing instruction of the cortex by the thalamus in any way directly contradict evidence for early significant differences in the ventricular zone? We will argue that it does not; there is in fact, a single mechanism, the specificity of the ventricular zone having the primary role of directing specific thalamic ingrowth, allowing specific thalamic regions to subsequently direct the differentiation of other features of cortical structure and function. The thalamocortical axons are the first subcortical input to reach isocortex (De Carlos and O'Leary 1992; Miller et al. 1993). The primary sensory nuclei and the nonspecific thalamic nuclei provide a convenient way to divide up the isocortex into different areas, as an individual cortical area receives a stereotypical thalamic projection. If a primary thalamic nucleus is removed early in development, connectional specificity is maintained: virtually no new thalamic regions innervate the denervated thalamic area, despite the availability of direct access (Miller et al. 1991). The entire cortical column is reduced in number by the degeneration of late-generated cortical neurons, and layer 4 fails to differentiate (Windrem and Finlay 1991). The connectional architecture of the cortex is entirely altered, both intracortically and subcortically, as described previously (Kingsbury et al. 1995).

The thalamus depends heavily on the cortex for trophic support as well. How specific are the requirements for support? It is well known that, if a cortical area is removed after the connections between cortex and thalamus are established, the associated thalamic nucleus degenerates (Finlay and Pallas 1989). What if a cortical area is removed before any contact is made between thalamus and cortex? Using information about the precise timing of thalamocortical innervation, we made lesions of the posterior cortex before thalamocortical axons were in residence under the subplate (Niederer et al. 1995). Removal of a cortical area before thalamic innervation caused the proportional loss of the part of thalamus expected to project there (Fig. 2). Thalamic areas appeared to be unable to colonize new cortical areas and employ them for trophic support. This finding is somewhat surprising, in that such rerouting or compression of projections has been shown in many other regions, for example, the midbrain tectum (Pallas and Finlay 1989) or the thalamic rewiring experiments done by Sur et al. (1988).

The prior experiments address the tangential specificity of trophic thalamocortical relationships, but the thalamus also restricts its projections to the isocortex by lamina: thalamic axons project mainly to layer 4 within isocortex. Does the

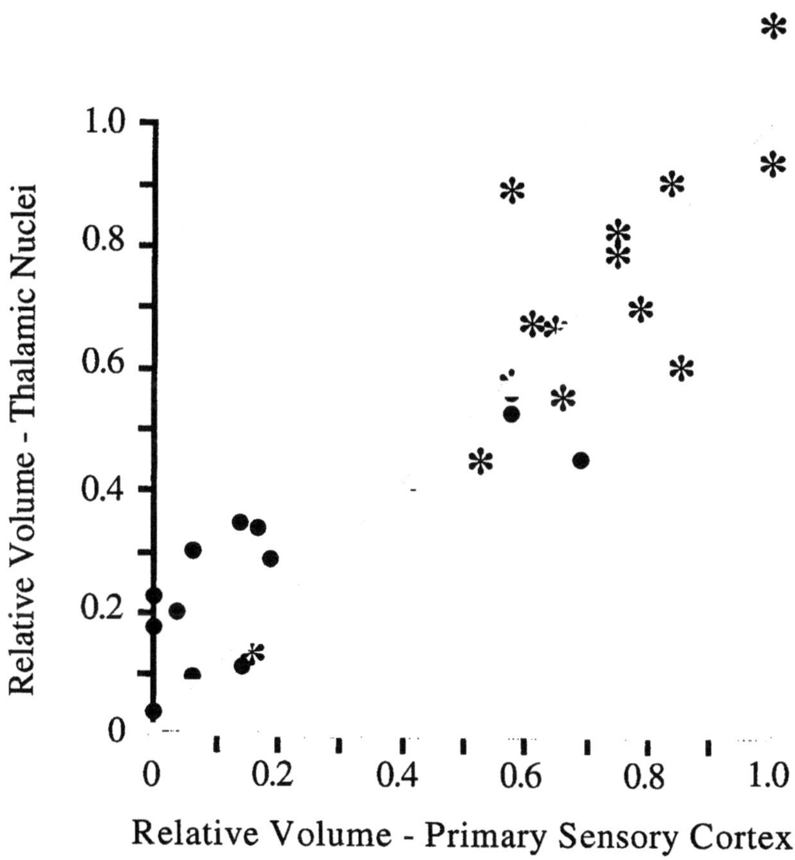

Fig. 2. Scatterplot showing the relationship between the relative volume of primary sensory cortex removed and its effect on the associated primary thalamic nucleus. Isocortex was removed on embryonic day 14 (two days before birth), before thalamic fibers have arrived. There is a direct relationship between the amount of isocortex removed and the degeneration of the associated thalamic nucleus

thalamus depend upon the cells to which it principally projects for either positional information or trophic support, or upon some other feature of the cortex? If layer 4 and the supragranular layers are almost entirely depleted in development with the mitotoxin MAM (Fig. 3 a), thalamic axons still project appropriately in the tangential plane (Woo and Finlay 1996; Woo et al. 1996). Loss of thalamic neurons is not proportional to the loss of the target zone; complete loss of cells with layer 4 morphology and up to 60% of the supragranular layers is necessary before cell loss is detectable in the thalamic nuclei afferent to the depleted cortex (Fig. 3 b). Overall, therefore, the particular nuclei of the thalamus are highly dependent upon their appropriate region of the cortex for trophic support but are surprisingly independent of the cortical layer they normally project to.

Development of the Thalamocortical Pathway

We have used the carbocyanine dyes DiI and DiA to explore the timing and behavior of both the thalamocortical and corticothalamic pathways. As others have also noted, we found that hamster thalamocortical axons seem to know their target and do not make exuberant projections that are later refined, at least at the level of cortical areas. Bayer and Altman (1991) have shown that secondary cortical areas are invaded by their thalamic projections before primary cortical areas, further demonstrating that thalamocortical projections do not simply look for unoccupied cortical areas but grow across uninnervated cortex and go directly to their correct targets.

Cortical target cells are thus exceptionally choosy as to the precise thalamic axons that they will accept, or thalamic axons are very choosy about what they innervate, suggesting that the thalamocortical system has a very specialized recognition system. Yet, according to the retroviral and culture experiments, individual cortical neurons are free to range large distances from their birthplace (O'Rourke et al. 1992; Walsh and Cepko 1990, 1992, 1993). A hypothesis to connect these apparently conflicting observations is that tangential migration may be an internal strategy of the cortex to designate "spheres of influence" within a given region or modality. If local differentiation occurs in the ventricular zone, this region might generate particular marker molecules that signal to thalamic or intracortical connections regions of varying levels of specificity, either region-specific or modality-specific. Tangentially migrating cells could mark the intermediate zone, the site of the thalamocortical axon growth, and build up a molecular matrix. For example, if a cell were specified to produce a particular marker in the ventricular zone, but migrated tangentially a distance of a few radial units before it entered the cortical plate, it would leave a chemical trail within the intermediate zone. Normally distributed migration would produce core regions with a large amount of the marker, whereas further away the chemical marker would be reduced and overlap with other locally generated markers. A single focus within a primary sensory area might attract the topographically organized pro-

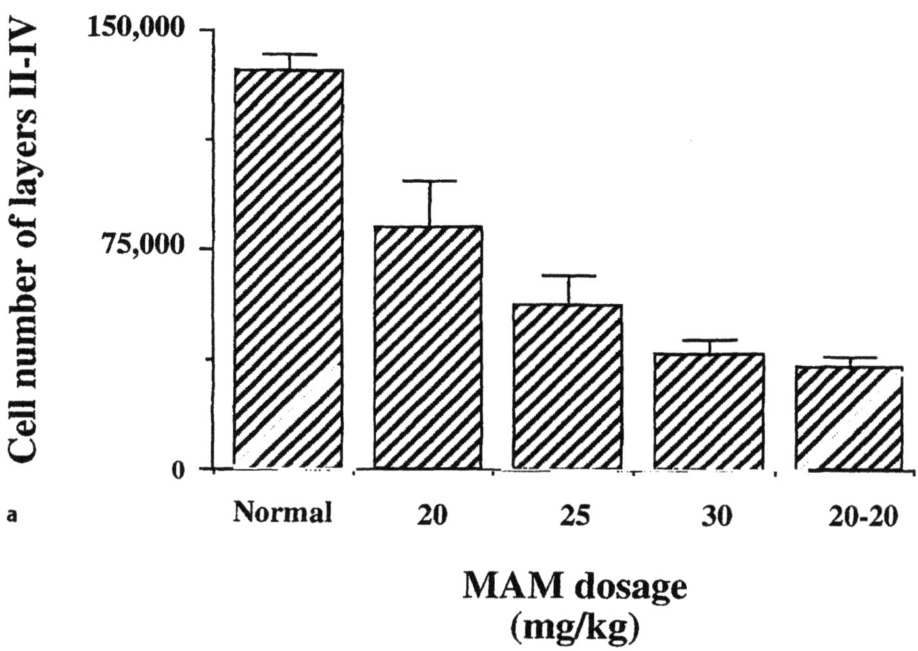

a

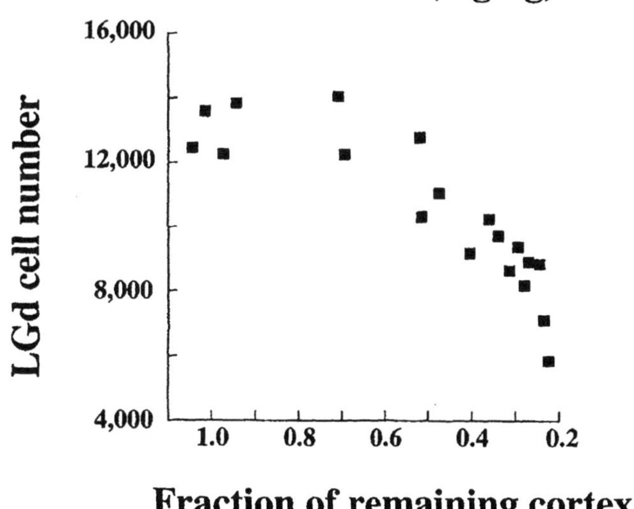

b

Fig. 3. a. Effect of prenatal MAM (methylazyoxymethanol) dosage on cell number of upper cortical layers (2–4). **b.** Scatterplot showing the relationship between thalamic cell number (y axis) and the fraction of remaining cortex (layers 2–4, x axis) after prenatal treatment with MAM. Depletion has to reach a critical point (59 %) before thalamic volume is affected. LGd, dorsal lateral geniculate

jection from the main sensory nucleus, whereas for areas around the primary region, several different projections or nonspecific projections might converge.

Transplantation and Coculture Studies may not be Incompatible with these Results

Coculture studies, as mentioned above, have shown that thalamocortical axons enthusiastically invade layer 4 no matter what the circumstances. A transplant may be considered an in vivo coculture study. The host has already provided the local regional information needed for thalamic innervation. In fact, the timing of these transplantation studies is such that the thalamic axons are already waiting underneath in the intermediate zone for the new tissue. Whether the transplanted cortex is actually able to make its normal intracortical projections is a more interesting question. Recent research has shown that, when embryonic visual cortex is transplanted to the somatosensory area, it does not make the subcortical connections appropriate to its new location. In fact, the transplanted cortex actually forms projections typical of visual cortex, its origin (Ebrahimi-Gaillard et al. 1994). Stanfield has done behavioral experiments on the transplanted animals and has found that the animals do not have complete behavioral function when the transplanted tissue is of visual cortex origin, but they do recover completely if the transplant is homotopic (Barth and Stanfield 1994). These experiments indicate that, even though transplanted tissue may show some of the features of its new location, the ectopic tissue has also retained properties of its initial location. Within our laboratory, we have shown that without thalamic input, intracortical connections do not become refined (Kingsbury et al. 1995). Connectivity has many facets; initial axon outgrowth can minimally vary in onset, rate, branchedness and target. It seems quite plausible that some of these features might be fixed in the ventricular zone, and others by thalamic instruction. Thalamic instruction may be particularly critical for activity-dependent sorting, regression and short-range construction of circuitry.

Consistent Inconsistencies of the Isocortex

Isocortex is specified early in development as to its appropriate modality, but intracortical connections are dependent upon input for proper development. Cells migrate radially but also tangentially. The laminar fate and number of cortical cells are controlled in the ventricular zone but also in the cortical plate. Thalamic projections show exquisite specificity, whereas corticofugal projections are promiscuous. Clearly the restriction of cortical fate is a serial process and arises from multiple information sources.

The radial dimension of the cortex is controlled by cellular mechanisms within the ventricular zone (McConnell 1988; McConnell and Kaznowski 1991). This leaves the tangential dimension, a simple two-axis array, already laid out in the ventricular zone and faithfully transposed to the cortical plate by radial glial cells (Götz 1995). Either or both could serve as sources of positional information that build up a subcortical matrix of positional information that directs the terminations of thalamic axons. Thalamic input in turn, at least through activity-dependent processes and possibly through other trophic or genetic instruction, further directs the cellular composition of the cortex and many aspects of its connectivity.

References

Angevine JB, Sidman RL (1961) Autoradiographic study of cell migration during histogenesis of cerebral cortex in the mouse. Nature 192:766–768

Bagnard D, Mann F, Henke-Fahle S, Bolz J (1995) Developmental mechanisms underlying the segregation and efferent cortical projections. Soc Neurosci Abstr 21:1285

Barbe MF, Levitt P (1991) The early commitment of fetal neurons to limbic cortex. J Neurosci 11:519–533

Barbe MF, Levitt P (1992) Attraction of specific thalamic input by cerebral grafts depend on the molecular identity of the implant. Proc Natl Acad Sci USA 89:3706–3710

Barth TM, Stanfield BB (1994) Homotopic, but not heterotopic, fetal cortical transplants can result in functional sparing following neonatal damage to the frontal cortex of rats. Cereb Cort 4:271–278

Bayer SA, Altman J (1991) Neocortical Development. New York, Raven

Bicknese AR, Sheppard AM, O'Leary DDM, Pearlman AL (1994) Thalamocortical axons preferentially extend along a chondroitin sulfate proteoglycan-enriched pathway coincident with the neocortical subplate and distinct from the efferent path. J Neurosci 14:3500–3510

Callaway EM, Katz LC (1990) Emergence and refinement of clustered horizontal connections in the cat striate cortex. J Neurosci 10:1134–1153

Cohen-Tannoudji M, Babinet C, Wassef M (1994) Early determination of a mouse somatosensory cortex marker. Nature 368:460–463

De Carlos JA, O'Leary DDM (1992) Growth and targeting of subplate axons and establishment of major cortical pathways. J Neurosci 12:1194–1211

Dehay C, Kennedy H, Smart IHM, Berland M (1991) Cell production gradients in the developing primate striate cortex. J Physiol (London) 434:13P

Ebrahimi-Gaillard A, Guitet J, Garnier C, Roger M (1994) Topographic distribution of efferent fibers originating from homotopic or heterotopic transplants – heterotopically transplanted neurons retain some of the developmental characteristics corresponding to their site of origin. Dev Brain Res 77:271–283

Finlay BL (1992) Cell death and the creation of regional differences in neuronal numbers. J Neurobiol 23:1159–1171

Finlay BL, Pallas SL (1989) Control of cell number in the developing mammalian visual system. Prog Neurobiol 23:207–234

Finlay BL, Slattery M (1983) Local differences in amount of early cell death in neocortex predict adult local specializations. Science 219:1349–1351

Ghosh A, Shatz CJ (1992) Pathfinding and target selection by developing geniculocortical axons. J Neurosci 12:39–55

Ghosh A, Shatz CJ (1993) A role for subplate neurons in the patterning of connections from thalamus to neocortex. Development 117:1031–1047

Ghosh A, Antonini A, McConnell SK, Shatz CJ (1990) Requirement for subplate neurons in the formation of thalamocortical connections. Nature 347:179–181

Götz M (1995) Getting there and being there in the cerebral cortex. Experientia 51:301–316

Johnston JG, van der Kooy D (1989) Protooncogene expression identifies a transient columnar organization of the forebrain within the late embryonic ventricular zone. Proc Natl Acad Sci USA 86:1066–1070

Kennedy H, Dehay C (1993) Cortical specification of mice and men. Cereb Cortex 3:171–186
Kingsbury MA, Miller B, Finlay BL (1995) Increased subcortical and intracortical projections after early thalamic ablation in the hamster. Soc Neurosci Abstr 21:45
Levitt P, Rakic P (1980) Immunoperoxidase localization of glial fibrillary acidic protein in radial glial cells and astrocytes of the developing Rhesus monkey brain. J Comp Neurol 193:817–848
Levitt P, Cooper ML, Rakic P (1981) Coexistence of neuronal and glial precursor cells in the cerebral ventricular zone of the fetal monkey. An ultrastructural immunoreactive analysis. J Neurosci 1:27–39
Levitt P, Cooper ML, Rakic P (1983) Early divergence and changing proportions of neuronal and glial precursor cells in the primate cerebral ventricular zone. Develop Biol 96:472–484
Luskin MB, Shatz CJ (1985) Studies of the earliest generated cells of the cat's visual cortex. J Neurosci 5:1062–1075
Luskin MB, Pearlman AL, Sanes JR (1988) Cell lineage in the cerebral cortex of the mouse studied in vivo and in vitro with a recombinant retrovirus. Neuron 1:635–647
Marin-Padilla M (1971) Early prenatal ontogenesis of the cerebral cortex (neocortex) of the cat (Felis domestica). A Golgi study. I. The primordial neocortical organization. Z Anat Entwicklungsgesch 134:117–145
McConnell SK (1985) Migration and differentiation of cerebral cortical neurons after transplantation into the brains of ferrets. Science 229:1268–1271
McConnell SK (1988) Fates of visual cortical neurons in the ferret after isochronic and heterochronic transplantation. J Neurosci 8:945–974
McConnell SK (1989) The determination of neuronal fate in the cerebral cortex. Trends Neurosci 342:12
McConnell SK, Kaznowski CE (1991) Cell cycle dependence of laminar determination in developing neocortex. Science 254:282–285
Miller B, Chou L, Finlay B (1993) The early development of thalamocortical and corticothalamic projections. J Comp Neurol 335:16–41
Miller B, Windrem MS, Finlay BL (1991) Thalamic ablations and neocortical development: alterations in thalamic and callosal connectivity. Cereb Cort 1:241–261
Misson JP, Austin CP, Takahashi T, Cepko CL, Caviness VS (1991) The alignment of migrating neural cells in relation to the murine neopallial radial glial fiber system. Cereb Cort 1:221–229
Molnár Z, Blakemore C (1995) Guidance of thalamocortical innervation. In: Bock GR, Cardew G (eds.). Development of the cerebral cortex, Vol. 193. Chichester, England. John Wiley & Sons Ltd, pp. 127–149
Niederer JK, Maimon G, Finlay BL (1995) Failure to reroute or compress thalamocortical projections after prenatal posterior cortex ablations. Soc Neurosci Abstr 21:2025
O'Leary DDM (1989) Do cortical areas emerge from a protocortex? Trends Neurosci 12:400–406
O'Leary DDM, Schlaggar BL, Tuttle R (1994) Specification of neocortical areas and thalamocortical connections. Ann Rev Neurosci 17:419–439
O'Rourke N, Dailey M, Smith S, McConnell S (1992) Diverse migratory pathways in the developing cerebral cortex. Science 258:299–302
Pallas SL, Finlay BL (1989) Conservation of receptive-field properties of superior colliculus cells after developmental rearrangements of retinal input. Visual Neurosci 2:121–135
Pallas SL, Sur M (1993) Visual projections induced into the auditory pathway of ferrets. 2. Corticocortical connections of primary auditory cortex. J Comp Neurol 337:317–333
Polleux F, Dehay C, Kennedy H (1995) Study of the early regionalisation of the ventricular zone and its role in the production of distinct neocortical areas. Soc Neurosci Abstr 21:1513
Price JL, Thurlow L (1988) Cell lineage in the rat cerebral cortex: a study using retroviral-mediated gene transfer. Development 104:473–482
Rakic P (1971) Guidance of neurons migrating to the fetal monkey neocortex. Brain Res 33:471–476
Rakic P (1972) Mode of cell migration to the superficial layers of fetal monkey noecortex. J Comp Neurol 145:61–84
Rakic P (1978) Neuronal migration and contact guidance in primate telencephalon. Postgrad Med J 54:25–40

Rakic P (1988) Specification of cerebral cortical areas. Science 241:170–176
Rauschecker J, Tian B, Korte M, Egert U (1992) Crossmodal changes in the somatosensory vibrissa/barrel system of visually deprived animals. Proc Natl Acad Sci USA 88:5063–5067
Schlaggar BL, O'Leary DDM (1991) Potential of visual cortex to develop an array of functional units unique to somatosensory cortex. Science 252:1556–1560
Stanfield B, O'Leary D (1985) Fetal occipital cortical neurons transplanted to the rostral cortex can extend and maintain a pyramidal tract axon. Nature 313:135–137
Sur M, Garraghty PE, Roe AW (1988) Experimentally induced visual projections into auditory thalamus and cortex. Science 242:1437–1441
Walsh C, Cepko CL (1990) Cell lineage and cell migration in the developing cerebral cortex. Experientia 46:940–947
Walsh C, Cepko CL (1992) Widespread dispersion of neuronal clones across functional regions of the cerebral cortex. Science 255:434–440
Walsh C, Cepko CL (1993) Clonal dispersion in proliferative layers of developing cerebral cortex. Nature 362:632–635
Windrem MS, Finlay BL (1991) Thalamic ablations and neocortical development: alterations of cortical cytoarchitecture and cell number. Cereb Cort 1:230–240
Woo TW, Finlay BL (1996) Cortical target depletion and ingrowth of geniculocortical axons: implications for cortical specification. Cereb Cort 6:457–469
Woo TW, Niederer JK, Finlay BL (1996) Cortical target depletion and the developing lateral geniculate nucleus; implications for trophic dependence: Cereb Cort 6:446–456
Woodward WR, Chiaia N, Teyler TJ, Leong L, Coull BM (1990) Organization of cortical afferent and efferent pathways in the white matter of the rat visual system. Neuroscience 36:393–401
Woolsey T, Wann J (1976) Areal changes in mouse cortical barrel following vibrissal damage at different postnatal ages. J Comp Neurol 170:53–66

Toxicity of Plasticity. Lessons from a model of developmental learning disorder

A.M. Galaburda

It is generally recognized that brain plasticity works at least as a mechanism for recovery of function after brain injury. Thus, for instance, the recovery of speech after an aphasia-producing stroke is thought to reflect brain changes of reorganization with beneficial results. Plasticity is thought to exert a useful function in degenerative conditions as well as in injury, which may explain why symptoms of disorders such as Alzheimer's and Huntington's disease take a long time to become clinically visible after a clear latency following the onset of neuropathologic changes. The commonly silent growth of brain tumors in terms of cognitive dysfunction is also attributed to brain plasticity changes that serve to compensate for the slow loss attributable to the tumor. Finally, the often cognitively silent cortical injury after a difficult gestation or perinatal complications is thought to result from the unusually active brain plasticity that occurs early in life (Aoki et al. 1988; Cotman et al. 1993; Freed et al. 1985; Milner 1974; Rauschecker 1995; Schneider 1979; Zilles 1992). However, regarding early brain plasticity (Schneider 1979), and occasionally after brain pathology later in life (Cotman et al. 1993), there has been a debate as to the positive and negative effects of brain plasticity. In the present chapter I will present evidence from my laboratory concerning a particular example of brain plasticity after early cortical injury that brings with it negative cognitive effects. Moreover, I suggest that this example may serve as a model for developmental learning disorders in humans following relatively trivial cortical injury early in life, whereby the injury itself is less likely to produce the clinical findings than is the brain plasticity that results from the injury.

The learning disorder modeled in the present experimental preparation is developmental dyslexia, a problem with language development that is most obviously visible as a difficulty with learning written language, but that also affects spoken language (Benton 1975; Liberman et al. 1985; Vellutino 1987). It is usually defined in terms of a discrepancy between measured intelligence and written language achievement. Other definitions that do not take into consideration a discrepancy measure (Shaywitz et al. 1992) include a larger set of possible etiologies for the reading and writing disorder, which cannot be modeled easily. Instead, the discrepancy definition, which is adopted by my laboratory, is more easily fitted into a theory stating that the language disorder, written and spoken, results from abnormal development of knowledge about the sound structure of the language in affected individuals from soon after the time of birth, or earlier (Kuhl et

A. M. Galaburda / Y. Christen (Eds.)
Normal and Abnormal Development of Cortex
© Springer-Verlag Berlin Heidelberg New York 1997

al. 1992). The present model also assumes to be true contemporary statements from cognitive neuroscience that claim that cognitive processes, including developmental cognitive processes, are represented at multiple stages and along several pathways, from the time a sensory stimulus reaches the nervous system from a sensory organ until it can be related to previously stored experiences and processes deep in the mind/brain (Kosslyn et al. 1992; Ungerleider et al. 1994). Thus, early in processing by the pathway, sensory/perceptual representations take place, mainly unimodal, whereas late in processing by the pathway, representations are associative and multimodal. The early representations are close to the sensory input, say visual or auditory, whereas the later representations are many synapses away from the sensory input, say in the frontal and anterior temporal regions. Furthermore, the need for different pathways at these several stages reflects the problem of having to represent aspects of the sensory input that are too different from one another to be implemented in the same brain machinery. Thus, for instance in the visual system, there is a need to represent object properties such as shape and color along different pathways from location and orientation in space (Haxby et al. 1991).

The model also requires that there is continually free flowing information from high level stages (those far from the input) to low level stages (those that are close to the input) and *vice versa.* This free flow of information aids in processing, particularly when information present at one level is of insufficient quality or quantity to allow for efficient access to meaning. For instance, when sensory input is fragmented or ambiguous, access to meaning may be aided by top-down information (from high to low level stages) providing clues, say, about context, which allows for the elimination of "nonfitting" possibilities (Kosslyn et al. 1992). Likewise, more detailed information from lower levels, say after examination of a larger protion of a partially examined stimulus and passed to higher levels in a bottom-up fashion, may lead more accurately and quickly to the appropriate stored memories or actions. Although this refers to the on-line analysis of sensory stimuli and their path to meaning or action, the same mechanism is thought to support the development of knowledge. Therefore, for example, the development of knowledge regarding the sound structure of an individual's native language requires first the representation of the fundamental sounds that compose the sounds of the language, which in turn are utilized downstream, further from the peripheral organ, to represent the phonemes of the language. The same fundamental sounds may be used in other ways by other stations belonging to other systems, say for musical processing. Therefore, the phonemic representations become a uniquely linguistic adaptation of the ability to process those fundamental sounds. It is understood that the elemental sounds composing the phonemes are represented closer to the input than the phonemes themselves. It is also implied that the same bottom-up and top-down processes described above for nonlinguistic visual objects are active in the on-line and developmental interaction between environmental sounds and representations of phonemes, words, sentences, and their access to meaning.

An additional feature of development is that failure to sense or perceive during critical developmental times the fundamental sounds that are part of the phonemic structure of an individual's native language will result in a problem with the ability to represent the phonemes of the language (Tallal et al. 1993; but see Rayner et al. 1995). Furthermore, the present model assumes that a difficulty with representing the phonemes of the language, no matter what the cause but allegedly the result of the difficulty with representing the relevant fundamental sounds, will in turn result in difficulties with spoken and written language, which benefit from, if not depend closely on, the accurate representation of the phonology.

On the other hand, it is conceivable that a disorder that affects the brain centers representing and processing the phonology of a language directly could, in a developmentally top-down manner, alter the development of low-level centers representing and processing more fundamental sounds. This scenario is the opposite of that advanced by Tallal (Tallal 1981; Tallal et al. 1993, 1995, 1973), and proposes that lower centers develop abnormally as a result of abnormal development of higher centers and not that higher centers develop abnormally as a result of abnormal development of lower centers. This is an important issue, which is at the center of a debate about the nature of the problem in developmental dyslexia: Do dyslexics have phonological deficiencies because they have failed to represent fundamental sounds that are required for the proper representation of the phonology of the language? Or do they have problems in low level sound processing and representation because they have developmental language abnormalities affecting phonological representation and processing? The present model, among other goals, aims at answering this question. Interesting observations regarding the effects of developmental brain plasticity following early brain injury have emerged during the attempt to answer the above question.

Evidence for Low and High Level Disturbances in Dyslexics

The evidence for functional abnormalities at low and high levels in developmental dyslexia will not be reviewed exhaustively in order to get to the plasticity data within the space allowed. However, I will provide examples to show that dyslexics exhibit difficulties performing low level, sensory/perceptual tasks as well as high level, associative tasks – i.e., at multiple stages of processing – and that they fail with challenging stimuli that are either visual or auditory – i.e., anlong multiple pathways.

Low Level, Sensory/Perceptual Difficulties

Such difficulties are seen in at least two modalities: auditory and visual. In the auditory modality Paula Tallal and colleagues (Tallal 1981; Tallal et al. 1993, 1995, 1973) showed many years ago that children with developmental language impairment, which included dyslexics, exhibited problems with auditory discrimination, particularly when two consecutive sounds are presented in rapid succession. With an inter-stimulus interval of about 500 msecs the affected individual could no longer distinguish between sounds. Since many language-specific sounds may differ from one another by changes occurring in as short a time as 20 msecs, Tallal has argued that children having this difficulty would be incapable of establishing representations for many native phonemes. This psychophysical evidence is not capable of precisely localizing the source of the abnormality in the brain, and the observed response could be the result of abnormalities anywhere along the pathway between sensory-perception and motor output. Therefore, it is possible that the low level deficits illustrated in this study may be the result of top-down influences acting on lower auditory centers.

On the other hand, a more precise localization of the sensory/perceptual problem is possible using event related potentials, and this technique was used in dyslexic and control subjects to demonstrate that the former show a delay in the processing of rapidly changing visual stimuli after only 100 msecs have elapsed since the presentation of the stimuli have reached the brain (Livingstone et al. 1991). Thus, after presentation of rapidly flickering checkerboard patterns at low contrasts we showed a prolongation and distortion of the averaged first segment of the wave, which indicated dysfunction somewhere between the retina and the primary visual cortex. The abnormality was most clearly visible when the magnocellular pathway of the visual system, i.e., the pathway responsible for processing rapidly changing visual stimuli at low contrasts, was most taxed by the stimulus. There has been further confirmation of dysfunction of the magnocellular pathway of the visual system of dyslexics (Borsting et al. 1996; Chase et al. 1993; Lehmkuhle et al. 1993), although not all studies have confirmed this abnormality.

Evidence also exists to support the claim that dyslexics suffer from disordered high level cognitive functions implicating mostly language. Thus, there appear to be problems with attention, verbal memory, semantics, pragmatics, and syntax, all functions thought to be processed via systems located more anteriorly in the brain and far from sensory input (see, for instance, Murphy et al. 1988; Wolf et al. 1992).

Anatomical findings appear to accompany the functional abnormalities, at both low and high levels. Thus, my colleagues and I showed several types and unique totography of developmental abnormalities in dyslexic brains obtained at autopsy and processed for architectonic and histometric analyses. In the auditory system close to the input – the left medial geniculate nucleus (MGN) – dyslexic brains showed a paucity of large neurons and an excess of large neurons as compared to control brains (Galaburda et al. 1994). In the lateral geniculate nucleus (LGN), dyslexic brains showed a decrease in the average caliber of neurons of the magnocellular, but not the parvocellular, layers (reduction to as much as two thirds of the average neuronal surface area; Livingstone et al. 1991).

Functional anatomic abnormalities have recently been demonstrated using brain activation methods in living dyslexics, which implicate both low and high level brain processors. Thus, Guinevere Eden and her colleagues, using functional magnetic resonance imaging, showed abnormalities in activation of the magnocellular system-linked motion perception area MT, as well as other abnormalities both upstream and downstream from MT (Eden et al. 1996). Similarly, Paulesu and colleagues (1996) showed significant differences in the activation pattern between dyslexic and nondyslexic subjects implicating a large portion of the perisylvian cortex, including anterior cortices that are located far from sensory input and that are implicated in high level language processing, i. e., phonologic and syntactic processing. Furthermore, since 1979, my laboratory has reported histological abnormalities of development affecting perisylvian cortex of dyslexics, which are attributed to abnormal neuronal migration (for a review, see Galaburda 1993).

Neuronal Ectopias

The cortex of the forebrain of developmental dyslexics coming to post-mortem examination have been shown in my laboratory to be studded with small nests of abnormally migrated neurons and glia in the molecular layer, which we have termed neuronal ectopias (or molecular layer ectopias). These anomalies represent abnormally migrated neurons and glia, which appear to reach the molecular layer via a rupture of the external glial limiting membrane during neuronal migration some time between 16 and 20 weeks of gestation (Sherman et al. 1992). While control brains seldom show even one of these anomalies (Kaufmann et al. 1989), dyslexic brains show large numbers, which are usually located in perisylvian cortex, including the cognitively important inferior prefrontal region far from sensory input. We have seen anomalies in primary auditory cortex, but not uniformly in this location, and no anomalies in primary visual cortex. The abnormalities are predominantly of the neuronal ectopia type, but focal microgyria is also seen. In female brains we have reported small myelinated glial scars comparable in size and location to those of the neuronal ectopia (for a discussion of the relationship between ectopias and myelinated scars, see Humphreys et al. 1990). We have also seen a preponderance of these anomalies in the left hemisphere compared to the right hemisphere of dyslexic brains.

In summary, anatomic anomalies are present in dyslexic brains both at high and low level sites *vis à vis* proximity to sensory inputs. The abnormalities at low level sites are predominantly in the size of neurons comprising the visual and auditory thalamic relay nuclei. The abnormalities at high level sites are disorders of neuronal migration, either molecular layer ectopias or focal microgyria, which affect perisylvian cortex, including areas far removed from sensory input and fundamentally multimodal in function. These abnormalities combined are apt to explain both sensory/perceptual and linguistic/cognitive disabilities in this clinical population.

Developmental Relationship Between High and Low Level Anomalies

My laboratory became interested in the question of whether the low level abnormalities, i.e., the small cells in the auditory and visual thalami, and the high level anomalies, i.e., the neuronal ectopias and microgyria affecting perisylvian cortex, were developmentally related, and whether one could be the result of the other. The anatomic answer to this question could help resolve the question of whether functional abnormalities in dyslexics begin as low level or as high level abnormalities, as discussed above. We knew that the high level abnormalities, i.e., the ectopias, had a fixed developmental window of origin, namely the period of neuronal migration to the neocortex. On the other hand, the changes in cell size could occur later as a result of modification of functional and anatomic inputs and outputs relating to the thalamic nuclei (see Wallace et al. 1992). Nevertheless, it was not possible to state with certainty that the ectopias preceded the histometric changes in the thalamus; nor was it possible to say that the latter were causally related to the former. Therefore, we developed an experimental animal model to test the hypothesis that events leading to the formation of cortical ectopias, or the ectopias themselves, occur earlier and are causally related to changes in cell sizes in connectionally related structures, including the thalamic nuclei. The accompanying functional hypothesis was that high level processing anomalies in dyslexics cause the low level sensory/perceptual difficulties rather than the other way around.

Migration Anomalies in the Experimental Model

We have been aware of a strain of mouse mutants that exhibit molecular layer ectopias, which my colleague Gordon Sherman has studied extensively in my laboratory (Sherman et al. 1990). Although a laborious process, the genetic basis for this abnormality, an autosomal recessive trait, is responding to mapping studies (Sherman et al. 1994). Furthermore, my colleague Glenn Rosen has developed a model for induction of cortical migration anomalies comparable to those seen in the dyslexic and mutant mouse brains (Humphreys et al. 1991; Rosen et al. 1992 a, b). He accomplishes this by injuring the developing cortex of a newborn rat or mouse with a freezing probe, which leads to the desired neuronal migration anomalies if the injury takes place before the period of neuronal migration has expired. On the other hand, injury after the completion of neuronal migration produces glial scars similar to those seen in the dyslexic female brains. Rosen has also shown that the duration and severity of the injury during neuronal migration determines whether a neuronal ectopia or a focus of microgyria will result (manuscript submitted for publication).

Functional and Anatomic Consequences of Induced Neuronal Migration Anomalies

A collaboration between Holly Fitch, then working in Paula Tallal's laboratory, and Glenn Rosen lead to a surprising discovery (Fitch et al. 1994). Rosen induced two sites of focal microgyria in the frontal lobes of newborn rats, one in each hemisphere, and Fitch tested the animals as adults in a specially designed apparatus for testing the animals' ability to perceive rapidly changing sounds. They found that, as compared to sham operated animals, the animals with microgyria lost their ability to respond to sound transitions for stimuli of short durations, a result that was reminiscent of the finding in developmentally language impaired children (Tallal et al. 1973). This represented a demonstration that the induction of such neuronal migration anomalies was capable of slowing processing of sounds, even though the anomalies were produced in the frontal lobes, far away from the auditory input.

The microgyria, as predicted by the hypothesis, was accompanied by a decrease in the sizes of neurons in the medial geniculate nucleus. In the initial batch of microgyric animals, the changes were seen only in the left MGN, with an excess of small neurons and a paucity of large neurons, just as in the human dyslexic brains (Herman et al. 1995). When additional animals were added to the initial batch, however, this lateralized finding disappeared, and the shifts in neuronal sizes were seen in both hemispheres. In either case, an important finding of this study was that, because the animals were otherwise normal until induction of the neuronal migration anomalies, the induction or the anomalies themselves were responsible for the changes in thalamic cell sizes. Therefore, it was also possible that the changes in the sizes of thalamic neurons of the dyslexic brains could result from the injury that lead to the formation of foci of abnormal neuronal migration in the cortex, or even to the abnormal foci of malformations themselves.

Injury and the resultant microgyria did not in any way involve the thalamic nuclei directly or cortical areas directly connected to them. Instead, the injury and malformations involved prefrontal cortex, which is only indirectly connected to the MGN. Connectional studies of these malformations have disclosed that the cortex in which they are located is abnormally connected (Jenner et al. 1995). A variety of tracing studies with anomalies at different sites show that anomalous connections may be present in the afferent and efferent systems. These connections may include preserved transient connections and *de novo* connections never seen during normal cortical development and considered to be the result of abnormal sprouting (Rosen et al., manuscript in preparation). In my view all of these connections, normal and anomalous, constitute the pathways by which malformation-induced changes can propagate to directly or indirectly connected cell groups and thus affect the development of connected neurons.

Toxicity of Plasticity

During behavioral testing, it was noted that microgyric female rats performed better than treated males, and an analysis of the results showed that the females, in fact, did not slow down after induction of microgyria (Herman et al., submitted for publication). Formal analysis of the behavioral results confirmed the suspicion that exposure to surgery did not produce abnormalities in sound processing in the females at high processing speeds, but it did so in the males. Anatomical analysis for possible differences in the size and location of the induced mi crogyria did not reveal any differences between the male and female lesions. In fact, female microgyria were in every way as large and affected the same regions as male microgyria. This indicated that, at least in the female case, the microgyria were not sufficient to produce the temporal processing abnormalities. Further anatomical analysis disclosed that, whereas in the males the microgyria were associated with changes in neuronal sizes in the MGN, as described above, no changes in cell sizes could be demonstrated in the female MGNs.

Conclusions

It is possible to regard the changes in the thalamus of early injured rats as plastic changes in response either to the microgyria themselves or to the events that cause the microgyria. However, it is clear that this plastic response occuring at a distance does not take place uniformly in all treated animals. Moreover, there may be asymmetric responses by the two hemispheres to the induction of bilateral malformations, although the results in this regard are at present inconclusive and need to be further examined. Thus, females do not respond comparably to males by exhibiting changes in neuronal sizes of thalamic nuclei as a consequence of events leading to cortical microgyria. Similarly, females, which show as severe bilateral focal microgyria as males, do not exhibit deterioration in their ability to process rapidly changing sounds as do males. It follows, therefore, that the changes in behavior are not related to the microgyria themselves but may instead be related to the secondary changes in the thalamus, closer to the sensory input channels. Absence of cell-size related plasticity in these peripheral channels is, in the females, associated with normal behavior. It is possible, however, that the females exhibit a different, more adaptive form of plasticity that does not translate into cell size changes in the thalamus. Thus far, it appears that a form of plasticity assessed by measuring neuronal sizes several synapses away from the malformation is associated with functional deficits, hence the phrase "the toxicity of plasticity." In addition, this research suggests that in the dyslexic brain it is also possible to entertain the notion that cortical malformations, albeit minor in extent, are in a position to alter the development of normally and abnormally connected neuronal groups, which in turn may be responsible for some or all of the functional deficits associated with this developmental learning disorder. A tempting addendum to these conclusions is that in human females, as seen in female rats, and perhaps in the right hemisphere, response to early cortical injury may be limited or of a different nature, such that behavioral disorders are less likely to follow. This may explain the age-old observation that developmental learning disorders in general, dyslexia in particular, affect females less often than males and affect the left hemisphere more than the right.

References

Aoki C, Siekevitz P (1988) Plasticity in brain development. Sci Am 259:56–64

Benton AL (1975) Developmental dyslexia: neurological aspects. Adv Neurol 7:1–47

Borsting E, Ridder WH, Dudeck K, Kelley C, Maksui L, Motoyama (1996) The presence of a magnocellular defect depends on the type of dyslexia. Vision Res 36:1047–1053

Chase C, Jenner AR (1993) Magnocellular visual deficits affect temporal processing of dyslexics. In: Tallal P, Galaburda AM, Llinas RR, von Euler C (ed) Temporal information processing in the nervous system, with special reference to dyslexia and dysphasia. New York Academy of Sciences, New York, pp 326–329

Cotman CW, Pike CJ, Cummings BJ (1993) Adaptive versus pathological plasticity. Possible contributions to age-related dementia. Adv Neurol 59:35–45

Eden GF, Van Meter JW, Rumsey JM, Maisog JM, Woods RP, Zeffiro TA (1996) Abnormal processing of visual motion in dyslexia revealed by functional brain imaging. Nature 382:66–69

Fitch RH, Tallal P, Brown CP, Galaburda AM, Rosen GD (1994) Induced microgyria and auditory temporal precessing in rats: A model for language impairment? Cereb Cortex 4:260–270

Freed WJ, de Medinacelli, Wyatt RT (1985) Promoting functional plasticity in the damaged nervous system. Science 227:1544–1522

Galaburda AM (1993) Neuroanatomic basis of developmental dyslexia. Behav Neurol 11:161–173

Galaburda AM, Menard MT, Rosen GD (1994) Evidence for aberrant auditory anatomy in developmental dyslexia. Proc Natl Acad Sci USA 91:8010–8013

Haxby JV, Grady CL, Horwitz B, Ungerleider LG, Mishkin M, Carson RE, Herscovitch P, Schapiro MS, Rapoport SI (1991) Dissocitation of object and spatial visual processing pathways in human extrastriate cortex. Proc Natl Acad Sci USA 88:1621–1625

Herman AE, Fitch RH, Galaburda AM, Rosen GD (1995) Induced microgyria and its effects on cell size, cell number, and cell packing density in the medial geniculate nucleus. Soc Neurosci Abstr 21:1711

Humphreys P, Kaufmann WE, Galaburda AM (1990) Developmental dyslexia in women: neuropathological findings in three patients. Ann Neurol 28:727–738

Humphreys P, Rosen GD, Press DM, Sherman GF, Galaburda AM (1991) Freezing lesions of the newborn rat brain: a model for cerebrocortical microgyria. J Neuropath Exp Neurol 50:145–160

Jenner AR, Galaburda AM, Sherman GF (1995) Connectivity of cortical ectopias in autoimmune mice. Soc Neurosci Abstr 21:1712

Kaufmann WE, Galaburda AM (1989) Cerebrocortical microdysgenesis in neurologically normal subjects: a histopathologic study. Neurology 39:238–244

Kosslyn SM, Koenig O (1992) Wet mind – the new cognitive neuroscience. The Free Press, New York

Kuhl PK, Kuhl PK, Williams KA, Lacerda F, Stevens KN, Lindblom B (1992) Linguistic experience alters phonetic perception in infants by 6 months of age. Science 255:606–608

Lehmkuhle S, Garzia RP, Turner L, Hash T, Baro TA (1993) A defective visual pathway in children with reading disability. N Engl J Med 328:989–996

Liberman IY, Shankweiler D (1985) Phonology and the problems of learning to read and write. Rem Spec Educ 6:8–17

Livingstone M, Rosen GD, Drislane FW, Galaburda AM (1991) Physiological and anatomical evidence for a magnocellular defect in developmental dyslexia. Proc Natl Acad Sci USA 88:7943–7947

Milner B (1994) Functional recovery after lesions of the nervous system. 3. Developmental process in neural plasticity. Sparing of language functions after early unilateral brain damage. Neurosci Res Prog Bull 12:213–217

Murphy LA, Pollatsek A, Well AD (1988) Developmental dyslexia and word retrieval deficits. Brain Lang 35:1–23

Paulesu E, Frith U, Snowling M, Gallagher A, Morton J, Frackowiak RS (1996) Is developmental dyslexia a disconnection syndrome? Evidence from PET scanning. Brain 119:143–157

Rauschecker JP (1995) Compensatory plasticity and sensory substitution in the cerebral cortex. Trends Neurosci 18:36–43

Rayner K, Pollatsek A, Bilsky AB (1995) Can a temporal processing deficit account for dyslexia. Psychonom Bull Rev 2:501–507

Rosen GD, Press DM, Sherman GF, Galaburda AM (1992 a) The development of induced cerebrocortical microgyria in the rat. J Neuropathol Exp Neurol 51:601–611

Rosen GD, Sherman GF, Richman JM, Stone LV, Galaburda AM (1992b) Induction of molecular layer ectopias by puncture wounds in newborn rats and mice. Dev Brain Res 67:285–291

Schneider GE (1979) Is it really better to have your brain lesion early? A revision of the "Kennard principle." Neuropsychologia 17:557–583

Shaywitz B, Fletcher JM, Holahan TM, Shaywitz SE (1992) Discrepancy compared to low achievement definitions of reading disability: results from the Connecticut longitudinal study. J Learn Dis 25:639–648

Sherman GF, Morrison L, Rosen GD, Behan PO, Galaburda AM (1990) Brain abnormalities in immune defective mice. Brain Res 532:25–33

Sherman GF, Rosen GD, Stone LV, Press DM, Galaburda AM (1992) The organization of radial glial fibers in spontaneous neocortical ectopias of newborn New-Zealand black mice. Dev Brain Res 67:279–283

Sherman GF, Stone LV, Doneberg VH, Beier DR (1994) A genetic analysis of neocortical ectopias in New Zealand Black mice. NeuroReport 5:721–724

Tallal P (1981) Language disabilities in children: perceptual correlates. Int J Pediatr Otorhinolaryngol 3:1–13

Tallal P, Piercy M (1973) Defects of non-verbal auditory perception in children with developmental aphasia. Nature 241:468–469

Tallal P, Miller S, Fitch RH (1993) Neurobiological basis of speech: a case for the preeminence of temporal processing. In: Tallal P, Galaburda AM, Llinas RR, von Euler C (eds) Temporal information processing in the nervous system, with special reference to dyslexia and dysphasia. New York Academy of Sciences, New York, pp 27–47

Tallal P et al. (1995) Dyslexia update. Irish J Psychol 16:194–268

Ungerleider LG, Haxby JV (1994) 'What' and 'where' in the human brain. Curr Opin Neurobiol 4:157–165

Vellutino FR (1987) Dyslexia. Sci Am 256:34–41

Wallace CS, Kilman VL, Withers GS, Greenough WT (1992) Increases in dendritic length in occipital cortex after 4 days of differential housing in weanling rats. Behav Neural Biol 58:64–68

Wolf M, Obregon M (1992) Early naming deficits, developmental dyslexia, and a specific deficit hypothesis. Brain Lang 42:219–247

Zilles K (1992) Neuronal plasticity as an adaptive property of the central nervous system. Anatomischer Anzeiger 174:383–391

New Genetic Insights into Cerebral Cortical Development

J.G. Gleeson and C.A. Walsh

Abstract

The pace of our understanding of cortical development has increased in recent years due in part to the availability of improved genetic resources. Whereas studies of neuronal proliferation and cell fate determination have been principally advanced by the study of vertebrate homologues of *Drosophila* neuronal determination factors, our understanding of neuronal migration has advanced by the analysis of several mouse and human cortical migration mutants. Among these mutations, the human disorder called periventricular nodular heterotopia displays complete failure of a population of neurons to leave the proliferative region. Rather than a complete failure of migration, other mutations such as the *reeler* mouse show abnormalities of targeting of the migrating neurons, associated with a failure of the normal interaction between two cortical neuronal populations, the preplate and the cortical plate. Several naturally occurring or engineered mouse mutants share a similar-appearing cortex, suggesting that multiple genes may be involved in a common pathway regulating preplate/cortical plate interactions. Two human disorders of neuronal migration, lissencephaly and the related disorder "double cortex" syndrome, bear some similarities to the mouse disorders and may also reflect abnormal preplate/cortical plate interactions. Through continued analysis of these mammalian cortical mutations, a clearer picture of the mechanisms underlying abnormal cortical migration is emerging.

Introduction

Several key advances in the study of cortical development have led to a better understanding of regulators orchestrating the growth and maturation of this incredible structure. One remarkable finding has been a growing recognition of extensive homologies and shared function for neurogenic genes in vertebrates and invertebrates. Second is the study of animal and human mutations affecting cerebral cortical development which lend insight into mechanisms guiding this development. Third is the identification of the genes responsible for two of these mutations affecting cerebral cortical development in animals and man. This chapter reviews normal cellular proliferation and migration in the developing cortex, introduces several animal and human mutants that affect neuronal migra-

A. M. Galaburda / Y. Christen (Eds.)
Normal and Abnormal Development of Cortex
© Springer-Verlag Berlin Heidelberg New York 1997

tion, reviews the genes that have been identified as being responsible for these disorders, and proposes a model of normal cortical cell migration based on recent findings.

Neuronal Proliferation

The cortex forms from a single, pseudostratified epithelium derived from the telencephalic bud, so that all cortical neurons are born beneath and must translocate to the cortical plate. Cortical neurons are born in the ventricular zone through a highly regulated process in which 11 rounds of cell division (in the mouse) give rise to the population of cells that constitute the adult cortex (Takahashi et al. 1995). Cell nuclei move within the radial cytoplasmic extension of the cell away from and back toward the ventricular surface by a process called interkinetic nuclear migration. DNA synthesis occurs during nuclear movement away from the ventricular surface, and cell division occurs when the nucleus has returned to the ventricular surface. In the process of cell division, a mitotic cleavage plane is established that can be horizontal, vertical, or intermediate with respect to the ventricular surface. Chenn and McConnell (1995) demonstrated that a vertical cleavage plane gives rise to two apparently equivalent daughter cells that then continue to proliferate. In contrast, a horizontal cleavage plane gives rise to a basal daughter that proliferates and an apical daughter that migrates away from the proliferative region towards the developing cortex.

Recent evidence suggests that cerebral cortical neurons share mechanisms of cell fate determination with *Drosophila*. In *Drosophila*, asymmetric localization of neuronal differentiation factors, such as Numb, to one of the two daughter cells during mitosis distinguishes the ultimate cellular identity of the two daughters, and in the mammalian cortex similar patterns are emerging. In the developing ferret brain, Notch 1, a mammalian homologue of Notch, is localized to the basal daughter of the horizontally oriented mitoses (Chenn and McConnell 1995), whereas Numb is also asymmetrically located but to the apical or apicolateral side of the mitotic cells (Zhong et al. 1996). In the case of symmetrical divisions, both daughters appear to inherit a similar quantity of Notch 1 and Numb proteins. These factors are likely essential in regulating cell proliferation and cell fate selection, since alterations in Notch expression cause changes in neuronal fate (Austin et al. 1995). In addition, mice homozygous for the HES-1 (*hairy* and *Enhancer of split* homologue-1) mutation, a helix-loop-helix factor that is thought to act downstream of Notch (Jarriault et al. 1995), exhibit severe neurulation defects and premature cortical neurogenesis (Ishibashi et al. 1995), whereas cell-autonomous over-expression of HES-1 using retroviral vectors blocks neurogenesis (Ishibashi et al. 1994).

Neuronal Migration to the Cortex

Recent work has provided a clearer understanding of the process of migration of postmitotic neurons to the cortex and a better understanding of genetic abnormalities of neuronal migration in both rodent and human. We will first discuss concepts of normal neuronal migration and then discuss disease models and how these models may shed light on the underlying molecular mechanisms governing this process.

Basic Structure of the Cortex

The adult mammalian cortex is composed of six layers of neurons. Layer I, present before the other neurons arrive, is the most superficial and is relatively cell sparse. Layers II–VI contain the bulk of cortical neurons and are structured in an orderly laminar pattern. These layers are formed in an orderly sequence, as demonstrated by ^{3}H-thymidine birthdating of neurons (Angevine and Sidman 1961). Layer VI neurons are generated first and migrate to their final position; layer V neurons are generated next and migrate past layer VI to reach their final position and so on, so that layer II represents the most recently generated neurons. Thus, the cortex is formed in an "inside-out" manner (Angevine and Sidman 1961; Rakic 1974). This implies that each successively generated neuron in the cortical plate must in turn bypass predecessors before ultimately settling at the outermost level of the cortical plate just below the marginal zone or layer I. Initially, the developing cortex is relatively thin as the first neurons are formed, so that initial neurons do not undergo extensive post-mitotic migration. In contrast, the majority of later-formed neurons must migrate increasingly larger distances to reach the cortex.

The Preplate is Comprised of Subplate and Marginal Zone Cells

There has recently been a greatly enhanced appreciation of the importance of the earliest-born cortical neurons in actually modulating the migration of later-born neurons. The oldest cortical neurons form a remarkably well-developed precocious structure, first termed the primordial plexiform layer (Marin-Padilla 1971) or preplate (Rickmann et al. 1977). As cellular proliferation continues and postmitotic neurons migrate out of the ventricular zone, a zone of densely packed cells appears in the middle of the preplate, and this new layer is termed the cortical plate (see Fig. 1). Marin-Padilla (1978) first proposed that the preplate is split into the marginal zone (MZ) and the subplate (SP) by later-formed neurons of the cortical plate. The cortical plate accumulates recently migrated neurons from the proliferative region and ultimately forms the bulk of the mature cortex. Since Marin-Padilla's original suggestion, ^{3}H-thymidine labeling experiments have confirmed and extended his model (Luskin and Shatz 1985). The precise

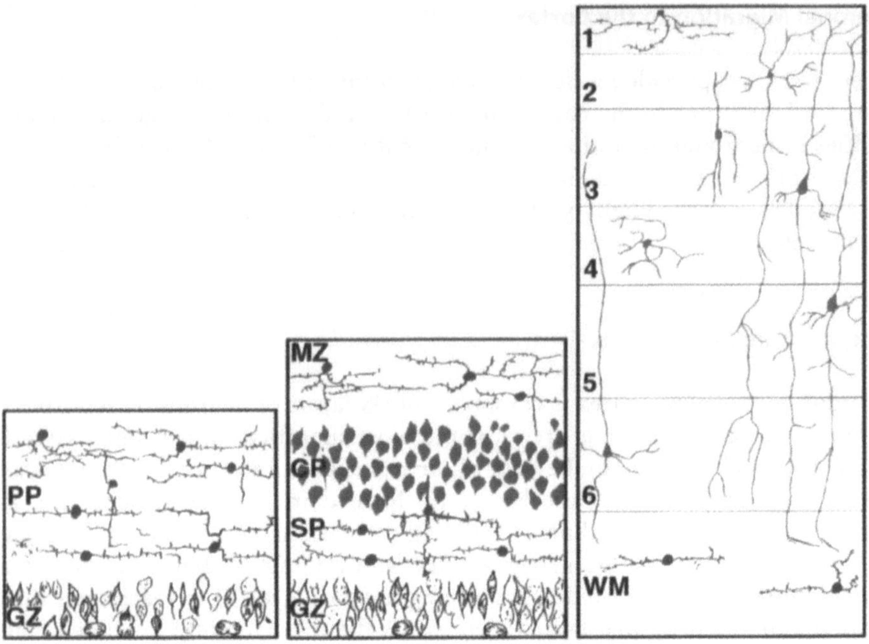

Fig. 1. A schematic of cerebral cortical development represented by (left) preplate stage, (middle) cortical plate stage, and (right) adult. These figures correspond roughly to E12, E16, and adult, respectively, in the rat. The cerebral cortex arises from progenitor cells in the germinal zone (GZ) along the ventricular surface. Early in neurogenesis, the earliest migrating neurons establish the preplate (PP). The eventual fate of these early-formed neurons of the PP is to form the Cajal-Retzius neurons of layer I, also known as the marginal zone (MZ), and the deep neurons of the subplate (SP). The preplate is split into the subplate and the marginal zone by the incoming cortical plate (CP) cells. As cortical plate neurons differentiate, a six layered structure is formed (right); marginal zone cells eventually form layer I, but very few of the subplate cells remain in the adult (seen in the white matter [WM]). Reproduced with permission from Reid and Walsh (1996)

fate of the cells that form the MZ and subplate is not completely clear; the MZ remnant in the adult cortex is layer 1 (Parnavelas and Edmunds 1983), but some of the Cajal-Retzius cells, the major cell type of the MZ, may not survive (Bradford et al. 1977). The subplate, while present in newborn animals, has only a remnant in later life, and some suggest that most or all subplate cells do not survive in the adult brain (Allendorfer and Shatz 1994).

Radial and Non-Radial Migration of Cortical Plate Neurons

Since the pioneering work of Rakic 25 years ago, based on Golgi impregnation and electron microscopy, there has been renewed interest in the central role of long, radially oriented glial cells in guiding postmitotic neurons from the ventricular zone to the cortex. Radial glial processes extend from the proliferative region to the pial surface (Rakic 1971 a, 1972), and migrating cortical plate neurons appear to be oriented in close contact with these fibers (Rakic 1971 b). The importance of radial glia, combined with the organization of the adult cortex into radially oriented functional units, or columns, gave rise to the additional suggestion that the functional columns of the cortex may represent clonally derived developmental units (Rakic 1974). However, studies over the last 10 years suggest that radial glia, while forming a crucial scaffolding that supports migrating neurons, do not exclusively determine the topographical location of those neurons, since a large proportion of neuronal migration can be observed to occur independent of and perpendicular to the radial glial system.

Evidence both for radial and non-radial migration began to emerge from retroviral studies in which infection of cortical progenitor cells produced labeling of clonal progeny. Clones of labeled cortical cells did not form tight radial columns, but instead occurred in loose cell clusters (Walsh and Cepko 1988), often composed exclusively of one cell type: oligodendrocytes, astrocytes, pyramidal neurons, or non-pyramidal neurons (Luskin et al. 1993; Parnavelas et al. 1991). There were additional observations that some migrating neurons were aligned perpendicular to radial glial fibers (O'Rourke et al. 1995). However, retroviral studies were limited in their ability to determine how far clonal sibling cells dispersed, because clones could not be identified independent of the pattern of cell migration in the cortex, i. e., sibling cells that migrated far apart could not be recognized as being sibling cells.

Retroviral libraries were developed as a means of tracing sibling cells independent of their patterns of migration, and studies with these libraries demonstrated surprisingly widespread dispersion of neurons derived from a single progenitor (Reid et al. 1995; Walsh and Cepko 1992). Daughters from a single progenitor may eventually reside not only in different cortical layers, but also in distinct functional domains, separated by up to millimeters in the adult rat brain. Analysis of chimeras (Fishell et al. 1990; Goldowitz 1987; Nakatsuji et al. 1991) and genetic mosaics formed by X-inactivation (Tan and Breen 1993) has also produced evidence both for radial and non-radial migration to the cortex, although the relative balance of the two modes of migration is not yet completely known.

Video microscopic studies have determined several types of non-radial migration of cortical neurons. Lateral migration occurs in the ventricular zone (Fishell et al. 1993) and as cells are migrating along radial glial fibers, although the relative contributions of ventricular zone and extra-ventricular zone migration towards total cell dispersion is unknown. Extra-ventricular zone lateral migration was documented in the ferret brain slice, where up to 17 % of neurons in the intermediate zone of neonatal ferret brain slices migrated with a trajectory of 30–60 degrees and right angle turns were seen in 5 % of radially migrating cells (O'Rourke et al. 1992). Given the curved geometry of the rapidly expanding cerebral cortex, it is possible that non-radial migration may allow greater cell dispersion or be required due to spatial constraints. Likewise, spatially regulated cues within the ventricular zone might influence the phenotypes of progeny cells produced by a single progenitor as it travels to different locations.

Is widespread clonal dispersion common to the development of the cortex of all species, or is it a phenomenon limited to the more "primitive" cortex of rodents? Retroviral cell labeling in the primate demonstrated some cell clusters with fairly precise radial organization, whereas other clonal patterns, termed "horizontal," constituted more loosely ordered cell groups. Unfortunately, this study did not use a retroviral library that would have allowed the issue of very widespread clonal dispersion to be addressed (Kornack and Rakic 1995). However, new studies in the ferret (Reid et al., 1997), a carnivore with a large, gyrencephalic, highly organized cortex, have shown extremely widespread dispersion of some neuronal clones across much of the cortex.

Widespread clonal dispersion may not affect all neuronal types in the cortex. Preliminary studies using direct cell labeling (O'Leary and Borngasser 1992) have suggested that the earliest-born neurons of the preplate may not show the same widespread dispersion that characterizes many neurons formed later in development. Thus, perhaps the clonal clustering of preplate cells may help the preplate encode positional information that may then be transmitted to the later-born cortical plate cells in a non-clonal fashion.

Functional Specification of Cerebral Cortical Regions

While proliferation and migration seem to occur mainly uniformly throughout the developing cortex, what are the factors that establish functional zones within the cortex? It is becoming clear that other *Drosophila* homologues may be important in specifying regional identity in the cortex. Specific families of genes that are homologues to *Drosophila* segmentation genes have been clearly implicated in encoding positional information in the hindbrain (Krumlauf et al. 1993). Distinct but related families of homeobox-encoding genes are also regionally expressed during mouse cortical development, and loss of function in one of these genes has been demonstrated to cause a severe cortical malformation. Emx1 and Emx2, related to the *Drosophila* gene *empty spiracles*, and Otx1 and Otx2, related to the *Drosophila* gene *orthodenticle*, are regionally expressed in the developing forebrain (Boncinelli et al. 1993) in a partially overlapping pattern. Otx2 is expressed in all dorsal and most ventral regions of telencephalon, diencephalon and mesencephalon; Emx1 expression is confined to the dorsal telencephalon, and Emx2 is expressed in dorsal telencephalon and small diencephalic regions, both dorsally and ventrally. Germline mutations in Emx2 have recently been demonstrated in some patients with severe schizencephaly, a disorder in which the cortex shows severe malformations and clefts (Brunelli et al. 1996). Therefore, cells arriving in the cortical plate may receive regional cues in the form of homeobox genes to establish anatomic boundaries and therefore possibly to determine functional fate.

Mutations Affecting Neuronal Migration to the Cerebral Cortex

Several cortical migration mutants, both human and rodent, show abnormal cellular migration, and it is beginning to become possible to think of some of these mutations as forming a genetic pathway governing neuronal migration. We will describe the mutants, and try to make an initial attempt at a model that is consistent with observations of cortical morphogenesis presented above.

Periventricular Nodular Heterotopia (PNH)

This human migrational abnormality is characterized by a normal appearing cortex, with a second layer of cortical neurons lining the subcortical ventricular surface, i.e., the remnant of the ventricular zone. The disease is often sporadic, occurring overwhelmingly in females, but recently has been found to be inherited in some pedigrees (Huttenlocher et al. 1994; Kamuro and Tenokuchi 1993). Affected females show seizures and sometimes mild mental retardation. Since affected individuals in these pedigrees were invariably females, and because of the excessive number of spontaneous abortions in these families, the disease was suggested to be X-linked dominant and likely lethal to males. This unusual pattern of inheritance is seen in several other disorders, including Aicardi syndrome and incontinentia pigmenti. X-linked dominant, male-lethal inheritance of PNH has been recently confirmed by direct linkage analysis (Eksioglu et al. 1996). The gene for PNH has been localized to Xq28, although the identity of the gene remains unknown. Linkage analysis narrows the candidate region to approximately seven megabases, although Xq28 is one of the most gene-rich regions known in humans, and the identification of the PNH gene may be long in coming.

Histopathological analysis has been performed on several brains with bilateral periventricular heterotopias (Adams et al. 1984; Eksioglu et al. 1996). The periventricular lesions fill the entire subependymal zone symmetrically beneath the cerebral cortex. The lesions consist of confluent nodules 2–10 mm in diameter. Microscopically, the nodules consist of highly differentiated neurons that look remarkably normal, except that they are oriented in multiple directions. Neuronal-glial relationships are indistinguishable from normal gray matter, multiple neuron types are evident, and the heterotopic neurons are richly innervated, although the source of this innervation is not clear (Eksioglu et al. 1996). It is also not known if the outer cortex shares reciprocal innervation with the heterotopic layer.

Given the interesting phenotype of PNH, one may form hypotheses of how loss of function of an X-linked gene gives rise to the phenotype. Because all females have two X chromosomes and males have one X chromosome, affected females will have one functional copy of the PNH gene and affected males will have no functional copies. Through Lyonization, neurons in the brains of females randomly inactivate one X chromosome during development (Tan et al. 1995).

Consequently, there will be two populations of differentiating neurons in the brain of PNH females: those that have inactivated the X chromosome with the diseased PNH gene and those that have inactivated the X chromosome with the normal PNH gene. Pathologically, affected females show two distinct populations of neurons: those that have migrated out of the proliferative region to form the normal cortex, and those that have remained in the proliferative region to form the heterotopia. Therefore, it is plausible that neurons that inactivate the normal X chromosome remain in the ventricular zone, whereas those that inactivate the diseased X chromosome have normal PNH gene function and migrate normally into the cortical plate. Supporting this hypothesis is the observation that PNH cells mature into adult-appearing neurons and assume the appearance of those neurons in the overlying cortex; it appears that the major abnormality of these neurons is their location. There are alternative hypotheses to explain the PNH phenotype that are consistent with Lyonization. These include the possibility that, instead of two populations of neurons, there are two populations of radial glial cells, which segregate neurons randomly. Other hypotheses that do not necessarily invoke Lyonization are excess proliferation or failure of apoptosis of a population of neurons that are secondarily unable to migrate.

The strikingly well-developed neurons in these nodules clearly indicate a major role for the PNH gene product in guiding human cerebral cortical development, whereas the gene is less likely to play a role in neuronal differentiation. The apparent segregation of neuronal subtypes within the PNH nodules (Eksioglu et al. 1996) bears intriguing similarity to the clustered cortical clones or subunits of widespread cortical clones in rodents and primates which often contain multiple cortical neurons of similar type or laminar location.

Disordered Interaction of Preplate and Cortical Plate in the Reeler Mouse

Over four decades ago, the mouse mutant *reeler* was identified, and in the ensuing years its phenotype has been described in great detail. The *reeler* mutant is caused by an autosomal recessive mutation located on mouse chromosome 5. *Reeler* homozygotes are unable to keep their hindquarters upright and frequently fall over on their sides when walking or running. Healthy *reeler* mice have fairly normal behavior except for difficulties in locomotion (Myers 1970). The initial observations in *reeler* were that migrating neurons came to populate densely and anomalously the normally sparsely populated cellular marginal zones of forebrain cortical structures, and that the positions of the major classes of neurons in the cerebral cortex were inverted (Caviness and Sidman 1973). A subsequent series of light, electron microscopic, and autoradiographic labeling studies in the 1970s and 1980s established that corresponding neuronal classes arise and execute their migrations out of the proliferative region according to an identical schedule in normal and mutant animals (Goffinet 1980; Pinto-Lord and Caviness 1979). However, migrating cortical cells in *reeler* fail to find their proper "addresses" within the dimension of the cortex.

Cortical development in the *reeler* mouse illustrates the interaction of the preplate and the cortical plate. The subplate has been intensively studied recently because of its apparent key role in orchestrating thalamocortical connection patterns (Allendorfer and Shatz 1994). In *reeler*, thalamocortical afferents take a bizarre course, bypassing their ultimate target neurons in layer IV, penetrating the entire thickness of the cortex, and running subpially in a layer termed the "superplate" before diving back down to contact layer IV cells (Caviness and Frost 1983). Since thalamocortical axons usually run in the subplate before entering the cortex, the axonal path in *reeler* first suggested that the subplate in *reeler* may actually occupy a subpial position, suggesting in turn that *reeler* mice may show a lack of the normal insertion of cortical plate neurons between preplate cells.

The pattern of lamination and mechanics of development in *reeler* have been clearly elucidated recently by an elegant series of studies by Ogawa et al. (1995). Using a variety of cell type-specific markers, Ogawa's group confirmed that the fundamental problem in *reeler* is a failure of splitting of the preplate. Secondly, they raised specific antisera and a monoclonal antibody, called CR-50, to an epitope that was present in normal mice and absent in *reeler* mice, and which presumably corresponds to the product of the *reeler* gene (D'Arcangelo et al. 1995). They showed that aggregates of normal cortical neurons will recapitulate the normal splitting of the preplate, whereas neurons from *reeler* mice will not. Finally, they showed that normal cells grown in the presence of the CR-50 monoclonal antibody show patterns of lamination similar to *reeler* cells.

The abnormalities of migration present in the *reeler* mouse may involve problems in the radial glial scaffold of the cerebral cortex. Pinto-Lord et al. (1982) found that the migrating neurons in the wild type are closely apposed to and coiled about the radial fiber until the final few micrometers of their migration, at which point they completely separate and move away from the radial fiber to their final position within the cortex. In the *reeler* mouse, however, the migrating neuron is blocked from completing its migration by other neurons that have incompletely separated from the glial fiber after terminating their migration. Moreover, a soluable unnamed factor produced in embryonic forebrain was found to be required for transformation of astrocytes to a radial glia (Hunter and Hatten 1995). However, radial glia in the *reeler* mouse show morphologic abnormalities, and although *reeler* produces this unnamed factor, the *reeler* astrocytes show a decreased ability to transform into radial glia in response to it (Hunter and Hatten 1996). These data suggest that the *reeler* gene may be involved in radial glia formation.

Through a fortunate transgene insertion, the *reeler* gene was identified recently by Miao et al. (1994), as was the corresponding protein named Reelin. Reelin mRNA was found to be localized to layer I in the mouse and expressed in high levels in Cajal-Retzius cells. Reelin resembles extracellular matrix proteins (D'Arcangelo et al. 1995) involved in cell adhesion. Its sequence implies that it is secreted. The amino terminus of Reelin is 25% identical to that of F-spondin, a protein that is secreted from the floor plate of the spinal cord and is thought to regulate the adhesion and extension of commissural axons (Klar et al. 1992).

There are eight EGF-like motifs, and they are most closely related to those of the extracellular matrix proteins tenascin C, tenascin X, restrictin and the integrin β-chain family. Therefore, Reelin may be a secreted factor important in cell-cell signaling and adhesion and is related to other adhesion molecules important in central nervous system development.

The mechanism by which later-born cells do not pass by earlier-born cells in *reeler* remains unclear. Somehow, although Reelin is expressed exclusively by Cajal-Retzius cells in the preplate and MZ (D'Arcangelo et al. 1995), migrating cortical cells that do not come in contact with the MZ show migrational abnormalities. Therefore, it may function as a diffusible factor or matrix protein. The function of Reelin will be further elucidated by identifying its target cell type, which may be the cortical plate cell, subplate cell, or radial glial fiber.

The *Scrambler* Gene May Represent Another Gene in the *Reeler* Pathway

A novel autosomal recessive mouse, discovered at The Jackson Labs in Bar Harbor, ME, displays a phenotype remarkably similar to *reeler*. Mice homozygous for the *scrambler* mutation display an unstable gait, trembling and ataxia similar to *reeler* mice. Histological examination of *scrambler* mutants show abnormalities similar to *reeler*. Both mutants show identical abnormal hippocampal cytoarchitectonics, with marked disruption of pyramidal and granule layers. Furthermore, *scrambler* neocortex shows poor lamination and an absence of the marginal zone, as do the *reeler* mutants (Sweet et al. 1996). The results of BrdU birthdating of neurons in the *scrambler* neocortex are strikingly similar to those observed in the *reeler* neocortex (Gonzalez et al. 1996). However, genetic analysis showed that *reeler* and *scrambler* are not allelic and map to different chromosomes (Sweet et al. 1996). It is tempting to postulate a common pathway that is interrupted in the *reeler* and *scrambler* mutants, although this is speculative.

Mutations in cdk-5 also May Generate a *Reeler*-Like Phenotype

The cdk-5 gene was isolated by virtue of its structural similarity to other cyclin-dependent kinases that are involved in cell cycle regulation; however cdk-5 is expressed mainly in postmitotic neurons of the CNS and PNS and does not have any known role in cell cycle regulation (Tsai et al. 1993). Instead, cdk-5 has been implicated in neuronal process extension, since dominant negative forms of cdk-5 inhibit neurite outgrowth in culture (Nikolic et al. 1996). Recently, homologous recombination in embryonic stem cells was used to engineer germline mutations in cdk-5, with a surprising phenotype (Ohshima et al. 1996). The homozygous cdk-5 mutants die at the day of birth and have widespread abnormalities in neuronal morphology and survival. However, the mutants also show a disorder of cortical lamination similar to the *reeler* phenotype, and like *reeler* the cdk-5 mutants also show abnormalities of cerebellar foliation and hippocampal architecture. The phenotype of the cdk-5 knockout needs to be more completely studied, but it is possible that cdk-5 may also represent a gene in the *reeler/scrambler* pathway.

The Double Cortex/X-Linked Lissencephaly Gene in Humans May also Involve Subplate-Cortical Plate Interactions

Double cortex (DC) is a remarkable disorder characterized by a band of heterotopic neurons located in the subcortical white matter, beneath a relatively normal cortex (Fig. 2). It has been termed subcortical band heterotopia or subcortical laminar heterotopia. Although it was described pathologically in 1893 (Matell 1893), it was first described radiologically in 1989, and this has led to better understanding of the disease. DC is usually sporadic but, like PNH, is seen almost exclusively (> 90 %) in females. The inheritance pattern of DC in two families (Pinard et al. 1994) is apparently X-linked dominant, with affected males displaying lissencephaly and affected females displaying DC. Lissencephaly is a much more severe neuronal migration disorder of the cortex.

Recent genetic evidence confirms that DC and X-LIS are mono-allelic, since the diseases are linked to the same locus within pedigrees displaying both diseases. Although the gene for DC remains unknown, DC/X-LIS has been linked to

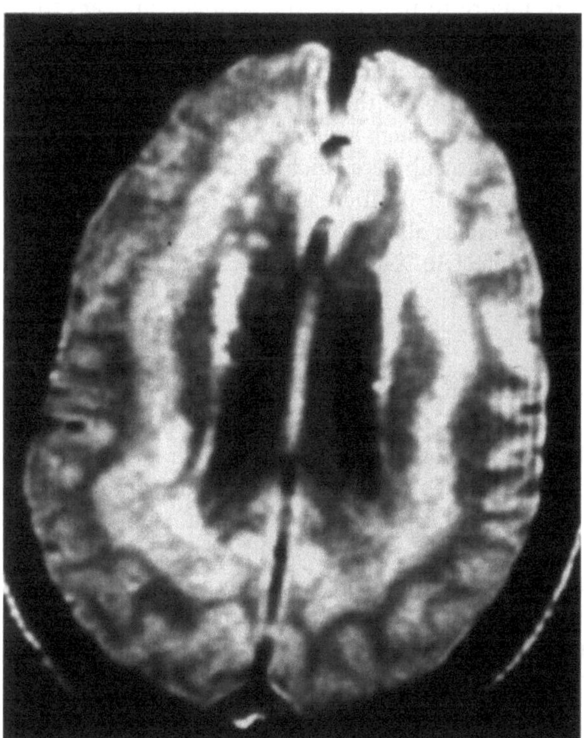

Fig. 2. MRI of double cortex. Outer cortex (light) and subcortical white matter (dark) appear normal. There is a second layer of subcortical heterotopic gray matter, taking the form of a band (light). The inner band of gray matter results from abnormal migration. Courtesy of Michael Berg, MD, University of Rochester

Xq21.3–q24 with a two-point LOD score of 3.3 (Gleeson et al. 1996) and a candidate region based upon recombination events of approximately 15 Mb. Although this is too large an area to evaluate candidate genes, a patient with lissencephaly who displays a *de novo* Xq22.3 balanced translocation may help narrow the region substantially (Dobyns et al. 1992).

Like PNH, affected females with DC likely have two genetically distinct populations of neurons that segregate based upon a single gene mutation. Neurons that inactivate the mutant X chromosome presumably migrate normally and form the relatively normal cortex, while neurons that inactivate the normal X chromosome have no remaining functional DC/X-LIS gene and consequently show migrational arrest beneath the normal cortex. Males with the DC/X-LIS mutation inherit no normally functioning DC genes and therefore essentially all neurons migrate abnormally, resulting in lissencephaly. Thus DC may be a cell-autonomous defect, although this is unproven.

A remarkable feature about the DC/X-LIS brain is that the heterotopic neurons migrate out of the ventricular zone but arrest roughly halfway to the cerebral cortex; why there? The heterotopic neurons are positioned near the subplate region, and in fact the heterotopic neurons may arrest because they fail to penetrate the subplate, while normally migrating neurons penetrate the subplate and enter the cortical plate. Several observations support this idea. First, the subplate is the only identifiable structure in the vicinity of the heterotopic neurons at the time of their migration. Second, failure to penetrate the subplate appears to be the essential defect in *reeler* and *scrambler* mice, and thus there is precedent for this as a cause of neuronal migration disorders. Third, the extent of the regional distribution of the band mirrors the location of the subplate during development: in regions where the subplate is prominent, such as lateral neocortex, heterotopic neurons are plentiful. Where there is no subplate, such as beneath the cingulate or temporal cortex (Kostovic and Rakic 1990), there are no heterotopic neurons. According to this model, in affected males all cells would arrest beneath the subplate, which may be the primary defect in X-linked lissencephaly and in other genetic forms of lissencephaly as well.

Does Chromosome 17 Lissencephaly Involve Preplate-Cortical Plate Interactions?

The "Type 1" lissencephaly (lissos = smooth), seen in males with DC/X-LIS, is a human neuronal migration disorder in which there is no normal gyral pattern and no laminar structure to the cortex. Lissencephaly is clinically characterized by severe disorders of cognitive development and seizures. Grossly, the brain is unusually small and of low weight; the interhemispheric fissure and separation of the two cerebral hemispheres is present. In the classical form, there is complete absence of any gyri or sulci with the exception of the central sagittal sulcus. Often, there is development of some secondary gyri and sulci, but these gyri are rudimentary. In these cases, the term pachygyria is used. The corpus callosum is often absent, or present as a thin remnant (Kuchelmeister et al. 1993).

Table 1. Mapped and/or cloned single gene loci associated with abnormal neuronal migration into the cortex. At least seven single gene loci are involved, and many of them may represent steps in a single pathway. Several mouse and human genes are shown, but present evidence suggets that each of the loci is distinct from the others, since the mouse and human disorders do not map to obviously syntenic regions

DISEASE NAME	SPECIES	MODE OF INHERITANCE	CHROMOSOMAL LOCATION	SYNTENIC LOCATION	RESPONSIBLE GENE	MECHANISM OF DISEASE
PNH	human	X-linked dominant male lethal	Xq28	mouse X	unknown	PROPOSED: cell autonomous failure to leave the ventricular zone
Reeler	mouse	autosomal recessive	chromosome 5	human 7	Reelin	PROVEN: failure of cells to penetrate subplate
Scrambler	mouse	autosomal recessive	chromosome 4	human 1	unknown	PROPOSED: identical to *reeler*
Type 1 lis (17-lis)	human	presumed dominant	17q13	mouse 9	PAF acetylhydrolase	PROPOSED: failure of all cells to penetrate subplate
DC/X-LIS	human	X-linked dominant	Xq23	mouse X	unknown	PROPOSED: cell autonomous failure of a population of cells to penetrate subplate
TISH	rat	autosomal recessive	unknown		unknown	PROVEN: two proliferatve regions

Mutations on chromosome 17 had been associated with lissencephaly (17-LIS) several years before the recognition of the X-linked lissencephaly locus (Dobyns et al. 1991). Lissencephaly associated with microdeletions of chromosome 17 and somatic dysmorphic features is called Miller-Dieker syndrome. Most cases have been sporadic; children have severe mental retardation and for this reason do not reproduce. 17-LIS is a hemideletion syndrome (disease in one of the two copies of the gene leads to lissencephaly). One may infer from this that the disease would be autosomal dominant if transmitted.

Microscopically, the lissencephaly cortex has identifiable layers and some refer to it as having four layers, but there is not clear layering of cells in the same way as is present in the normal cortex. Instead, the layering is more structural, from superficial to deep: a molecular layer (layer I), an outer cellular layer (layer II), a sparsely cellular layer (layer III) and a thick inner cellular layer (layer IV). The narrow white matter typically shows myelin and often contains neuronal heterotopias. There is often an enlarged ventricular system, and very rudimentary olivary nuclei in the medulla (Kuchelmeister et al. 1993). The pathogenesis of the brain malformation is not known.

It is not clear how the layers in lissencephaly and in the normal cortex are related; neither is it clear how lissencephaly relates to other migrational abnormalities. The molecular layer is usually more cellular than that of the normal marginal zone, and there is no clear border representing the junction between the molecular layer and the deeper layers. Layer IV is by far the thickest layer, spanning several times the thickness of normal human cortex. This suggests that the normal cortical mantle is present as layer IV in lissencephaly.

The small deletions of chromosome 17 present in many children with lissencephaly led to the identification of a gene known as LIS-1, found to be deleted in most of these patients (Reiner et al. 1993). LIS-1 was found to be identical to a 45KD regulatory subunit of intracellular platelet-activating factor (PAF) acetylhydrolase (Hattori et al. 1994) and showed homology to the β-subunit of the G protein family (Reiner et al. 1993). PAF acetylhydrolase hydrolyzes PAF, a potent pro-inflammatory bioactive lipid. Mouse homologues of LIS-1 have been mapped to three different chromosomal loci, none of which corresponds to the *reeler* or *scrambler* loci. Two antibodies directed against the carboxy and amino terminal motifs of human LIS-1 have been generated (Mizuguchi et al. 1995) and show the LIS-1 gene product is restricted to the central nervous system. When examined in Miller-Dieker patients, one group found near complete absence of reactivity (Mizuguchi et al. 1995), whereas another found reactivity in this population of patients (Clark et al. 1996). Preliminary immunohistochemical staining with these antibodies in normal fetal brain showed staining in all cortical plate neurons after migration was complete; however, at early stages of cortical plate development, cortical staining was limited to the large neurons in layer I and subplate cells (Clark et al. 1996). Therefore, the LIS-1 gene product may be involved in the preplate/cortical plate interaction.

The mechanism by which LIS-1 causes lissencephaly is not clear, but it may be similar to *reeler*, in that lissencephaly may result from failure of splitting of the preplate by the cortical plate. This hypothesis is supported by two pieces of evidence. First, the cortex of X-LIS males is indistinguishable from typical lissencephaly microscopically, and in the X-LIS allelic disorder DC, the abnormally located neurons are arrested in the vicinity of the subplate. Second, the preferential localization of LIS-1 immunoreactivity to the preplate cells early in development supports this hypothesis. However, if the two lissencephalies are related pathogenetically to the *reeler* and *scrambler* mice, why does the cortex look so different in the mouse mutants versus the human mutants? Perhaps this difference is due to species differences, or perhaps it signifies different underlying mechanisms of abnormal migration. The true relationship of *reeler/scrambler* and lissencephaly/DC needs to be more fully investigated. As the underlying mechanism for abnormal cell migration in the two diseases is clarified, the relationship between these two diseases will become more clear.

TISH, or Double Cortex Rat

A novel autosomal recessive rat mutation was discovered in 1995 (Lee et al. 1995) which displays a phenotype of "double cortex" and has been named TISH, for telencephalic internal structural heterotopia. The heterotopic region is present throughout the neocortex but is not prominent in temporal areas. Mild to moderate ventriculomegaly is observed, and seizure activity is sometimes present in adult TISH animals. Otherwise, the external phenotype and behavior of the TISH animals are normal. Preliminary Golgi studies demonstrate the presence of neocortical-like pyramidal cells in the heterotopic region. However, these cells are not uniformly oriented in a vertical manner and their placement suggests that a strict laminar pattern is not present. In contrast, the overlying neocortex appears normal. Connectivity studies using retrograde Flurogold tracer indicate that the TISH region establishes afferent and efferent connections with cortical and subcortical sites. In addition, the findings are consistent with the idea that the TISH region represents a second cortical structure in which neocortical connections may establish a dual representation. Finally, preliminary neurogenesis studies have been presented that indicate that the period of neurogenesis of cells destined for the TISH region overlaps with the period of neocortical neurogenesis. In addition, the cells in the TISH region exhibit a laminar neurogenetic gradient that is similar to, but less strict than in the neocortex. The TISH region exhibits a rough inside-out gradient; however, the adult distribution of cells generated on a given embryonic day appears much less precise.

It is not clear at this point what relationship the TISH rat will have to the human "double cortex" disorder. Several features differentiate them. First, the TISH rat is an autosomal recessive disorder; there is no phenotype in the heterozygous state. In contrast, the human disorder DC is X-linked dominant, and our model of pathogenesis depends on this fact. Second, the TISH rat heterotopia is not clearly a band. In DC, the heterotopic neurons clearly take the form of a band located beneath the normal cortex. Third, preliminary data indicate the presence of two proliferative regions in the TISH rat (Collins and Lee 1996), so that there may be an entirely different mechanism from DC giving rise to the TISH phenotype. Taken together, the disorders may involve a distinct genetic pathway.

A Model for Cortical Migrational Abnormalities

Summarizing the above mutants of cortical migration, there are at least seven single gene mutations causing neuronal migration defects in the cortex as a mapped simple Mendelian locus (Table 1), and there are probably many more loci that have not yet been mapped. These include both human and rodent mutations; most of the human syndromes lack rodent counterparts, and vice versa. Why this is the case is not clear. Mouse *reeler* mutations have arisen spontaneously and repeatedly, yet there is no known human homologue of the *reeler* phenotype. Perhaps some rare human cases of lissencephaly are due to defects in the *reeler* or *scrambler* genes, although there has been no human *reeler* equivalent with the severe cerebellar abnormalities described to date.

Mutations affecting neuronal migration in the cortex appear to show both cell autonomous and non-cell autonomous actions. Whereas the Reelin protein appears to be a secreted signal and the disorder is non-cell autonomous (Mullen 1977), PNH and DC/X-LIS might be candidates for cell autonomous disorders since they have sharply defined phenotypes in a mosaic state. Because the human diseases discussed are X-linked, females are essentially mosaics with two populations of cells, and in this situation, there appears to be sharp segregation of these two cell populations. This leads to the intriguing disorders PNH and DC, both with a normal-appearing cortex and a subcortical collection of neurons presumably segregated on the basis of the responsible gene. This mechanism of action, however, remains to be proved.

It is clear that the splitting of the preplate by the incoming cortical plate is a critical event for the development of the cortex. Failure of the migrating cortical plate cells to split the preplate leads to *reeler* and possibly *scrambler* phenotypes in the mouse and may be the underlying mechanism in DC/X-LIS and chromosome 17 lissencephaly as well. In only two of the seven disorders listed in Table 1 has the responsible gene been identified to date – Reelin and LIS-1 – and it is not clear how these two very different proteins cause similar defects in neuronal migration. To be sure, as the mechanism of these two proteins is elucidated and genes and proteins for the other diseases are identified, we may begin to develop a better understanding of neuronal migration.

Acknowledgments

JGG is supported by the NINDS grant No. #5K12NS01701–04. CAW is supported by grants from the NINDS (#K08 NS01520, R01 NS32457), The Human Frontier Science Program and the Rita Allen Foundation. Michael Berg, MD, University of Rochester, supplied radiographic material.

References

Adams JH, Corellis JAN, Duchen LW (1984) Greenfield's neuropathology. 4th Edition. New York, John Wiley and Sons, pp 411–413

Allendorfer KL, Shatz CJ (1994) The subplate, a transient neocortical structure: its role in development of connections between thalamus and cortex. Ann Rev Neurosci 17:185–218

Angevine JB, Sidman RL (1961) Autoradiographic study of the cell migration during histogenesis of the cerebral cortex in the mouse. Nature 192:766–768

Austin CP, Feldman DE, Ida JA, Jr., Cepko CL (1995) Vertebrate retinal ganglion cells are selected from competent progenitors by the action of Notch. Development 121:3637–3650

Boncinelli E, Gulisano M, Broccoli V (1993) Emx and Otx homeobox genes in the developing mouse brain. J Neurobiol 24:1356–1366

Bradford R, Parnavelas JG, Lieberman AR (1977) Neurons in layer I of the developing occipital cortex of the rat. J Comp Neurol 176:121–132

Brunelli S, Faiella A, Capra V, Nigro V, Simeone A, Cama A, Boncinelli E (1996) Germline mutations in the homeobox gene EMX2 in patients with severe schizencephaly. Nature Genet 12:94–96

Caviness VS, Jr., Sidman RL (1973) Time of origin or corresponding cell classes in the cerebral cortex of normal and reeler mutant mice; an autoradiographic analysis. J Comp Neurol 148:141–151

Caviness VS, Jr., Frost DO (1983) Thalamocortical projections in the reeler mutant mouse. J Comp Neurol 219:182–202

Chenn A, McConnell SK (1995) Cleavage orientation and the asymmetric inheritance of Notch 1 immunoreactivity in mammalian neurogenesis. Cell 82:631–641

Clark G, Mizuguchi M, Swann J, Barnes J, Antalffy B, Armstrong D (1996) Localization of the Miller-Dieker lissencephaly gene product to the Cajal-Retzius cells. Ann Neurol (Abstr) 40:1996

Collins JL, Lee KS (1996) Cortical neurogenesis in the Tish mutant rat: Evidence for dual proliferative zones. Soc Neurosci Abstr 2:1013

D'Arcangelo G, Miao GG, Chenn SC, Soares HD, Morgan JI, Curran T (1995) A protein related to extracellular matrix proteins deleted in the mouse mutant reeler. Nature 374:719–723

Dobyns WB, Curry CJ, Hoyme HE, Turlington L, Ledbetter DH (1991) Clinical and molecular diagnosis of Miller-Dieker syndrome. Am J Human Genet 48:584–594

Dobyns WB, Elias ER, Newlin AC, Pagon RA, Ledbetter DH (1992) Causal heterogeneity in isolated lissencephaly. Neurology 42:1375–1388

Eksioglu YZ, Scheffer IE, Cardenas P, Knoll J, DiMario F, Ramsby G, Berg M, Kamuro K, Berkovic SF, Duyk GM, Parisi J, Huttenlocher PR, Walsh CA (1996) Periventricular heterotopia: an X-linked dominant epilepsy locus causing aberrant cerebral cortical development. Neuron 16:77–87

Fishell G, Rossant J, Vanderkooy D (1990) Neuronal lineages in chimeric mouse forebrain are segregated between compartments and in the rostrocaudal and radial planes. Dev Biol 141:70–83

Fishell G, Mason CA, Hatten ME (1993) Dispersion of neural progenitors within the germinal zones of the forebrain [published erratum appears in Nature, May 20, 1993;363:286]. Nature 362:636–638

Gleeson J, Allen K, Ross M, Scheffer I, Berkovic S, Berg M, Motte J, Walsh C (1996) Linkage analysis of "double cortex"/X-linked lissencephaly. Ann Neurol (Abstr) 40:290

Goffinet AM (1980) The cerebral cortex of the reeler mouse embryo. An electron microscopic analysis. Anat Embryol 159:199–210

Goldowitz D (1989) Cell partitioning and mixing in the formation of the CNS: Analysis of the cortical somatosensory barrels in chimeric mice. Dev Brain Res 35:1–9

Gonzalez JL, Russo C, Sweet H, Davisson M, Goldowitz D, Walsh C (1996) Neocortical birthdating analysis on *scrambler*, a novel mouse mutant, reveals a *reeler*-like phenotype. Soc Neurosci Abstr No 393.16, Vol 2:987

Hattori M, Adachi H, Tsujimoto M, Arai H, Inoue K (1994) Miller-Dieker lissencephaly gene encodes a subunit of brain platelet-activating factor acetylhydrolase [corrected] [published erratum appears in Nature Aug 4, 1994; 370:391]. Nature 370:216–218

Hunter KE, Hatten ME (1995) Radial glial cell transformation to astrocytes is bidirectional: regulation by a diffusible factor in the embryonic forebrain. Proc Natl Acad Sci USA 92:2061–2065

Hunter K, Hatten M (1996) A model for the differentiation and maintenance of the embryonic radial glial scaffold in mammalian forebrain. Soc Neurosci Abst No 22:1206

Huttenlocher PR, Taravath S, Mojtahedi S (1994) Periventricular heterotopia and epilepsy. Neurology 44:51–55

Ishibashi M, Moriyoshi K, Sasai Y, Shiota K, Nakanishi S, Kageyama R (1994) Persistent expression of helix-loop-helix factor HES-1 prevents mammalian neural differentiation in the central nervous system. EMBO 13:1799–1805

Ishibashi M, Ang SL, Shiota K, Nakanishi S, Kageyama R, Guillemot F (1995) Targeted disruption of mammalian hairy and Enhancer of split homolog-1 (HES-1) leads to up-regulation of neural helix-loop-helix factors, premature neurogenesis, and severe neural tube defects. Genes Devel 9:3136–3148

Jarriault S, Brou C, Logeat F, Schroeter EH, Kopan R, Israel A (1995) Signalling downstream of activated mammalian Notch. Nature 377:355–358

Kamuro K, Tenokuchi Y (1993) Familial periventricular nodular heterotopia. Brain Devel 15:237–241

Klar A, Baldassare M, Jessell TM (1992) F-spondin: a gene expressed at high levels in the floor plate encodes a secreted protein that promotes neural cell adhesion and neurite extension. Cell 69:95–110

Kornack DR, Rakic P (1995) Radial and horizontal deployment of clonally related cells in the primate neocortex: relationship to distinct mitotic lineages. Neuron 15:311–321

Kostovic I, Rakic P (1990) Developmental history of the transient subplate zone in the visual and somatosensory cortex of the macaque monkey and human brain. J Comp Neurol 297:441–470

Krumlauf R, Marshall H, Studer M, Nonchev S, Sham MH, Lumsden A (1993) Hox homeobox genes and regionalisation of the nervous system. J Neurobiol 24:1328–1340

Kuchelmeister K, Bergmann M, Gullotta F (1993) Neuropathology of lissencephalies. Childs Nervous System 9:394–9

Lee KS, Lanzino G, Schottler F, Collins J, Kassell N, Hiramatsu K, Goto Y, Hong S, Berr S, Caner H, Yamamoto H, Omary R, Okonkwo D, Jane J (1995) A novel rat with a double cortex: 1. General features. Soc Neurosci Absts No 26.12, Vol. 1:46

Luskin MB, Shatz CJ (1985) Studies on the earliest-generated cells in the cat's visual cortex: cogeneration of subplate and marginal zones. J Neurosci 5:1062–1075

Luskin MB, Parnavelas JG, Barfield JA (1993) Neurons, astrocytes, and oligodendrocytes of the rat cerebral cortex originate from separate progenitor cells: an ultrastructural analysis of clonally related cells. J Neurosci 13:1730–1750

Marin-Padilla M (1971) Early prenatal ontogenesis of the cerebral cortex (neocortex) of the cat (*Felis Domestica*). A Golgi study. I. The primordial neocortical organization. Z Anat Entwickl-Gesch 134:117–145

Marin-Padilla M (1978) Dual origin of the mammalian neocortex and evolution of the cortical plate. Anat Embryol 152:109–126

Matell M (1893) Heterotopie der grauen Substanz in den beiden Hemispharen des Grosshirns. Arch Psychiat Nervenkr 25:124–136

Miao GG, Smeyne RJ, D'Arcangelo G, Copeland NG, Jenkins NA, Morgan JI, Curran T (1994) Isolation of an allele of reeler by insertional mutagenesis. Proc Natl Acad Sci USA 91:11050–11054

Mizuguchi M, Takashima S, Kakita A, Yamada M, Ikeda K (1995) Lissencephaly gene product. Localization in the central nervous system and loss of immunoreactivity in Miller-Dieker syndrome. Am J Pathol 147:1142–1151

Mullen RJ (1977) Genetic dissection of the CNS with mutant-normal mouse and rat chimeras. In: Cowan WM, Ferrendeli JA (eds) Approaches to the cell biology of neurons. Vol 2. Bethesda, MD, Society for Neuroscience, pp 47–65

Myers W (1970) Some observations on reeler, a neuromuscular mutation in mice. Behav Genet 1:225–234

Nakatsuji N, Kadokawa Y, Suemori H (1991) Radial columnar patches in the chimeric cerebral cortex visualized by use of mouse embryonic stem cells expressing beta-galactosidase. Dev Growth Differ 33:571–578

Nikolic M, Dudek H, Kwon YT, Ramos YF, Tsai LH (1996) The cdk5/p35 kinase is essential for neurite outgrowth during neuronal differentiation. Genes Devel 10:816–825

Ogawa M, Miyata T, Nakajima K, Yagyu K, Seike M, Ikenaka K, Yamamoto H, Mikoshiba K (1995) The reeler gene-associated antigen on Cajal-Reztius neurons is a crucial molecule for laminar organization of cortical neurons. Neuron 14:899–912

Ohshima T, Ward J, Huh C-G, Longenecher G, Veeranna, Pant H, Brady R, Martin L, Kulkarni A (1996) Targeted disruption of the cyclin-dependent kinase 5 gene results in abnormal corticogenesis, neuronal pathology and perinatal death. Proc Natl Acad Sci USA 93:11173–11178

O'Leary DDM, Borngasser DJ (1992) Minimal dispersion of neuroepithelial cells during generation of the cortical preplate. Soc Neurosci Abst 18:925

O'Rourke NA, Dailey ME, Smith SJ, McConnell SK (1992) Diverse migratory pathways in the developing cerebral cortex. Science 258:299–302

O'Rourke NA, Sullivan DP, Kaznowski CE, Jacobs AA, McConnell SK (1995) Tangential migration of neurons in the developing cerebral cortex. Development 121:2165–2176

Parnavelas JG, Edmunds SM (1983) Further evidence that Retzius-Cajal cells transform to nonpyramidal neurons in the developing rat visual cortex. J Neurocytol 12:863–871

Parnavelas JG, Barfield JA, Franke E, Luskin MB (1991) Separate progenitor cells give rise to pyramidal and nonpyramidal neurons in the rat telencephalon. Cereb Cortex 1:463–468

Pinard J-M, Motte J, Chiron C, Brain R, Andermann E, Dulac O (1994) Subcortical laminar heterotopia and lissencephaly in two families: a single X-linked dominant gene. J Neurol Neurosurg Psych 57:914–920

Pinto-Lord MC, Caviness VS, Jr (1979) Determinants of cell shape and orientation: a comparative Golgi analysis of cell-axon interrelationship in the developing neocortex of normal and reeler mice. J Comp Neurol 187:49–69

Pinto-Lord MC, Evrard P, Caviness VS, Jr (1982) Obstructed neuronal migration along radial glial fibers in the neocortex of the reeler mouse: a Golgi-EM analysis. Brain Res 256:379–393

Rakic P (1971 a) Guidance of neurons migrating to the fetal monkey neocortex. Brain Res 33:471–476

Rakic P (1971 b) Neuron-glia relationship during granule cell migration in developing cerebellar cortex. A Golgi and electronmicroscopic study in Macacus Rhesus. J Comp Neurol 141:283–312

Rakic P (1972) Mode of cell migration to the superficial layers of fetal monkey neocortex. J Comp Neurol 145:61–83

Rakic P (1974) Neurons in rhesus monkey visual cortex: Systematic relation between time of origin and eventual disposition. Science 183:425–427

Reid CB, Liang I, Walsh C (1995) Systematic widespread clonal organization in cerebral cortex. Neuron 15:299–310

Reid CB, Tavazoie SF, Walsh CA (1997) Clonal dispersion and evidence for asymetric cell division in ferret cortex. Development 124 (12), 1441–50

Reid CB, Walsh CA (1996) Early development of the cerebral certex. In: Neural Development and Plasticity, Elsevier, Amsterdam, pp 17–30

Reiner O, Carrozzo R, Shen Y, Wehnert M, Faustinella F, Dobyns WB, Caskey CT, Ledbetter DH (1993) Isolation of a Miller-Dieker lissencephaly gene containing G protein β-subunit-like repeats. Nature 364:717–721

Rickmann M, Chronwall BM, Wolff JR (1977) On the development of non-pyramidal neurons and axons outside the cortical plate: the early marginal zone as a pallial anlage. Anat Embryol 151:285–307

Sweet HO, Bronson R, Johnson K, Cook S, Davisson MT (1996) Scrambler, a now neurological mutation of the mouse with abnormalities of neuronal migration. Mammalian Genome 7:798–802

Takahashi T, Nowakowski RS, Caviness VS, Jr (1995) The cell cycle of the pseudostratified ventricular epithelium of the embryonic murine cerebral wall. J Neurosci 15:6046–6057

Tans SS, Breen S (1993) Radial mosaicism and tangential cell dispersion both contribute to mouse neocortical development. Nature 362:638–640

Tan SS, Faulkner-Jones B, Breen SJ, Walsh M, Bertram JF, Reese BE (1995) Cell dispersion patterns in different cortical regions studied with an X-inactivated transgenic marker. Development 121:1029–1039

Tsai LH, Takahashi T, Caviness VS, Jr., Harlow E (1993) Activity and expression pattern of cyclin-dependent kinase 5 in the embryonic mouse nervous system. Development 119:1029–1040

Walsh CA, Cepko CL (1988) Clonally related neurons show several patterns of migration in cerebral cortex. Science 255:1342–1345

Walsh C, Cepko CL (1992) Widespread dispersion of neuronal clones across functional regions of the cerebral cortex. Science 225:434–440

Zhong W, Feder J, Jiang M, Jan L, Jan Y (1996) Asymmetric localization of a mammalian Numb homolog during mouse cortical neurogenesis. Neuron 17:43–53

Genetic and Environmental Determinants of Neocortical Development: Clinical Applications

P. Evrard, S. Marret and P. Gressens

Abstract

This chapter reviews the conceptual and methodological tools that permit us to analyze the influences of genetic, epigenetic and environmental determinants on developing neural tissue, including developing neocortex, during early life.

During the period of *preparation of the neural germinative epithelium*, environmental influences, including nutritional, circulatory, maternal and placental factors, can interfere with the genetic program in a very complex way. Whole postimplantation mouse embryo cultures are a powerful tool for the study of environmental, nutritional, hypoxic and genetic factors at this developmental phase. At this step we review particularly the role of recently recognized growth and neurotrophic factors [such as vasointestinal peptide (VIP)] on brain growth, an avenue to explore microcephaly and intrauterine growth retardation. At the step of *neuronal migration*, we review mainly the *alcohol and cocaine-induced disturbances of corticogenesis*. Alcohol and cocaine interfere with the development of the human fetal brain. The teratogenic mechanisms of these drugs on neurogenesis were recently explored in animal models. *Cocaine* severely disturbs neocortical architecture, disrupting horizontal and vertical lamination and inducing an abnormal array of the axonal-dendritic bundles. Cocaine also alters several steps of gliogenesis. The severity of malformations is variable but is evident in all animals exposed to doses comparable to the doses by human cocaine abusers. The cocaine-induced cortical pattern seems to be the result of dyschronologic mitoses and of a defect of the radial glial cells. This pattern may represent the pathological basis of the neuropsychological modifications described in *in utero* cocaine-exposed children. Immediate early genes (IEGs; *c-fos, c-jun, and zif*-268) are disturbed by cocaine during development. As transcription factors. IEGs can directly dysregulate target genes. *Ethanol* locally enhances cell death in the primitive neuroepithelium. During neuronal migration, ethanol induces a premature transformation of the radial glial guides into astrocytes. Ethanol also inhibits late gliogenesis. The resulting postmigratory neocortex displays an abnormal neuronal pattern almost completely deprived of vertical columnization. These glial-neuronal disturbances can explain neuropathological and clinical features of the fetal alcohol syndrome. *At the end of and after neuronal migration*, the infragranular layers are a sensitive target for perfusion failures/hypoxias around mid gestation. During the second half of pregnancy, *the trans-*

A. M. Galaburda / Y. Christen (Eds.)
Normal and Abnormal Development of Cortex
© Springer-Verlag Berlin Heidelberg New York 1997

formation of radial glial cells into astrocytic precursors is a target for environmental disturbances, among which nutritional factors and hypoxia/ischemia are candidates. Several *residual neuronal migrations or displacements* occurring after mid-gestation have complex pathophysiological relationships with ischemia/ hypoxia and with circulatory events. The late germinative zone produces migrating astrocytic precursors for the upper neocortex. If confirms the *dual origin of astrocytic precursors* and suggests the transitory existence of a late astroglial protomap destined to the upper cortex, which could explain cortical consequences of periventricular leucomalacias (PVLs) and intraventricular hemorrhages (IVHs).

The developing brain is weltering in a complex mixture including excitotoxic substances, cytokines, and growth and neurotrophic factors. These substances are sometimes environmental friends, like maternal VIP, which could prevent brain intrauterine growth retardation. They are sometimes excellent endogenous friends like neurotrophic excitatory agents in physiological conditions. They often become dangerous killers triggered by environmental signals, like hypoxias/ischemias and toxins produced by intrauterine infections launching the excitotoxic ascade. Deviant neopallial architectural patterns are produced by excitotoxins at the successive steps of early life. Using ibotenate (IBO), a competitive glutamatergic agonist, it is possible to reproduce most patterns of fetal and neonatal dysmorphogenetic and destructive lesions encountered in human developmental neuropathology. This model can also be used to assess scavengers and excitotoxin antidotes, such as magnesium and neurotrophic factors.

Introduction

This review is focused on conceptual and methodological tools that permit us to analyze the influences of several environmental and genetic determinants on developing neural tissue during early life. This overview of basic environmental and genetic mechanisms in developmental neuropathology is one of the contributions developed gradually over the last two decades by our group in partnership with other research groups. Several grevious review papers summarize the progression of our thinking in this field (Caviness et al. 1981; Evrard et al. 1984, 1989, 1992, 1993, 1995, 1997; Evrard and Minkowski 1989).

Preparation of the Neural Germinative Epithelium

The developmental period between the isolation of the neural groove and the onset of neuronal migration, or Phase II of neural development, covers the preparation of the neural germinative epithelium, with the neuronal and glial "lineaging," the emergence of a first protomap (Gressens et al. 1992 a, d; Caviness et al. 1981; Takahashi et al. 1993), and the appearance within the germinative layers of cell precursors destined to the prospective neocortical units (Gressens and Evrard 1993). This crucial "Phase II" step, which ends at embryonic day 11.5 in the mouse and at gestational week 9 in the human, is heavily modulated by envi-

Table 1. Whole postimplantation mouse embryo culture

Method (Gressens et al. 1992 a) 80 % human serum/20 % rat serum 5 % CO2/95 % air 5–26 somites
Pharmacological influences investigated in our laboratroy Cocaine: defective differentiation of radial glial cells (Gressens et al. 1992 b) Retinoic Acid: delayed neuronal-glial lineaging (Gressens et al. 1992a) Caffeine: modification of cell cycles in the germinative zone, with enhancement of mitotic and pycnotic indices, and acceleration of neural maturation (Marret et al. 1995 c) Ethanol: cell death in the ventromedial part of the mesencephalic-prosencephalic junction in the primitive neuroepithelium (Gressens et al. 1992c) Valproate: defect of closure of the neural tube (Gofflot et al. 1995) Ibotenate (excitotoxin): no effect (Marret et al. 1995 d) Role of the cellular receptor tissue factor(TF) in embryonic blood vessel development (Carmeliet etal. 1996)

Modified, from Evrard et al. 1995

Table 2. Neopallial units

Perivascular neuronal units (Kuban and Gilles 1985)
Domains coupled by gap junctions (Yutes et al. 1992)
Protomap (Rakic 1971)
Dispersed clones (Walsch and Cepko 1992; Austin and Cepko 1990)
The compromise (Tan and Breen 1993)
The pyramidal and local circuit (Marin-Padilla 1990)
Vertical units deriving from the radial glial fascicles (Evrard et al.)
Vertical physiological column (Montcastle 1957)
Variants of szentagothai (Eccles)
Neuropsychological units (Changeux) (Edelman)

From Gressens and Evrard 1993

ronmental influences that can be assessed on whole postimplantation mouse embryo cultures, a powerful tool for teratogenic investigations (Gressens et al. 1992 b, c; Tables 1 and 2). The application of vasoactive intestinal peptide (VIP) to these whole cultured mouse embryos led to the discovery that VIP is a crucial determinant of CNS growth (Gressens et al. 1993); in vivo, extraembryonic VIP could be the source of this regulating factor of embryonic and CNS growth (Gressens et al. 1994; Hill et al. 1996). VIP could be involved in intrauterine growth retardation. It stresses the importance of maternal and/or placental environment and the influence of circulatory factors. Whole mouse embryo cultures (Gressens et al. 1992 a, 1993) were submitted to 10^{-7} to 10^{-3} ibotenate (IBO), a glutamatergic agent, and cultured for 24 hours in a mixture of human and rat sera from the 7-somite stage to the 20–21-somite stages (see below). During this developmental period of neural tube closure and early premigratory differentiation of the neuroepithelium, no macroscopic or light-microscopic effect of this glutamatergic agent has been detected (Marret et al. 1995 a, b).

Energy Supply for Migrating Neurons

Migrating neurons, often distant from blood vessels, seem to use mainly the anaerobic glycolysis. They are in close contact with radial glial cells (RGCs), which are stuffed with particulate glycogen, whereas all normal neuronal elements are devoid of any detectable particulate glycogen. RGCs extend from the ventricular zone (close to the choroid plexuses and vessels) to the pial surface, an area richly irrigated by the epiparenchymal leptomeningeal plexus. The RGCs, fully and uniformly loaded with particulate glycogen, seem to act as energy corridors (Kuban and Gilles 1985; Gadisseux and Evrard 1985). Disturbances of neuronal migration due to cocaine and alcohol, fetal alcohol and fetal cocaine syndromes have been described in several laboratories, including ours (Fig. 1; Gressens et al. 1992 b, c).

We have no clear information concerning the mechanism of glucose and/or lactate transfer from glial glycogen to the migrating neuron. The developing neurons metabolically seem to transform from a predominantly anaerobic condition when migrating towards a more aerobic condition after their postmigratory settlement in a more vascularized cortical plate.

The Postmigratory Infragranular Layers: A Sensitive Target for Perfusion Failures/Hypoxias between 18 and 28 Gestational Weeks

The basic lesion in postmigratory polymicrogyria[1] is a postmigratory laminar depopulation predominating in the infragranular part of the cortical plate and especially in the neocortical layer V–IV a (Barkovich et al. 1992; Dvorak and Feit

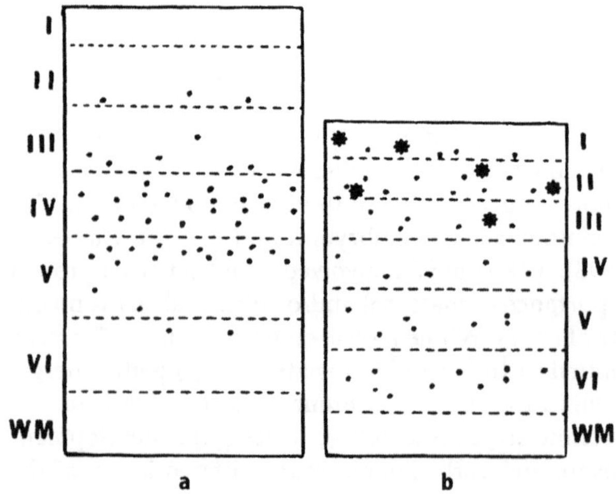

Fig. 1. Cocaine-induced disturbances of corticogenesis. Camera lucida drawings of the distribution of neurons (*) and glial cells (*) intensely labeled with [3H]thymidine in dorsolateral frontal cortex of the mouse. [3H]thymidine was injected into pregnant dams at E14, and pups were sacrificed on P5. **A**, control animal. **B**, cocaine-treated animal. The layer IV–V predominance of neural distribution in controls is contrasted with a random laminar distribution evident in drug-exposed mice. WM, white matter; I to VI, neocortical layers. (From Gressens et al. 1992 b)

[1] There are several types of microgyria, discussed in Evrard et al. 1989.

Table 3. Hypoxias, Ischemias, Perfusion, Failures: a few types and pathophysiological mechanisms[1]

1st trimester	Middle trimester	3rd trimester
Environment of neural epithelium	Laminar hypoxia	Matrix hemorrhages
	Migratory corridors	Leukoencephalopathies (before 32 weeks)
	Micropolygyrias (second half of middle trimester)	Glial migration (peak: 26 weeks ?)
	Neuronal escapes	Porencephaly Multicystic encephalomalacias Ulegyrias

[1]From Evrard et al. (1995)

1977; Evrard et al. 1994; Gressens et al. 1991; Richman et al. 1974). Clinicopathologic correlations, including sequential ultrasound, from our European Multicenter Study confirm that perfusion failures leading to layered polymicrogyrias often happen after the end of neuronal migration and before the peak of gyration (between 20 and 28 weeks of gestation; Evrard 1988). Some cases of microgyria originate earlier, before the end of neuronal migration, and display late migratory and laminar necrosis or depopulation. These variants of microgyria have been reviewed by Evrard et al. (1984). Using intracerebral injections of the excitotoxin IBO in developing mouse brain, we recently created a model of neocortical excitotoxic destruction (see below). When killing the postmigratory neurons, IBO mimics remarkably well the hypoxic aggression. The comparative chronology of angiogenesis, neuronal metabolism and neuronal migration provides a framework to interpret the pathophysiology of the microgyria and of the different types of brain hypoxia at the successive developmental stages (Table 3). The influence of IBO at the consecutive neurodevelopmental steps is detailed below.

Neuronal Displacements Outside of Migratory Corridors

Superficial ectopias, due to late neuronal escape through holes in the limiting pial-glial membrane, settle in the plexiform zone or in the leptomeningeal space. They are a constant feature of fetal alcohol syndrome and are encountered in cases of developmental microcephaly, mental retardation of unknown origin, and localized injuries to the basal membrane or to the surface of developing brain. Marin-Padilla (1990, 1993) studied the occurrence of neuronal ectopias in injuries to the cortical surface acquired late in gestation and in early premature newborns, associated with pial-glial membrane ruptures caused by infectious, vascular, ischemic and/or hypoxic agents. He demonstrated the influence of these ectopias on the surrounding normotopic neurons.

As alluded to earlier, after neuronal destruction by ibotenic acid (see below) and local freezing, neuronal displacements outside of the migratory corridors or along preserved RGCs are sometimes conspicuous enough to refill a destroyed layer with a full complement of neurons. In these neuronal compensations, the role of an adaptation of programmed cell death (apoptosis) is under consideration.

The Dual Origin of Astrocytic Precursors

Most neurons destined for the human neocortex are produced before 20 weeks of gestation. The Perinatal Project of the National Institutes of Health provided a serially processed material allowing Jammes and Gilles (1983) to demonstrate that the germinative zone reaches its maximum volume around the 26th week of intrauterine life. The conspicuous germinative zone in premature infants produces oligoglial precursors and most astrocytic precursors destined for the upper half of the neocortex. These data have implications concerning the mechanisms of brain disturbance associated with *germinal matrix-intraventricular hemor-*

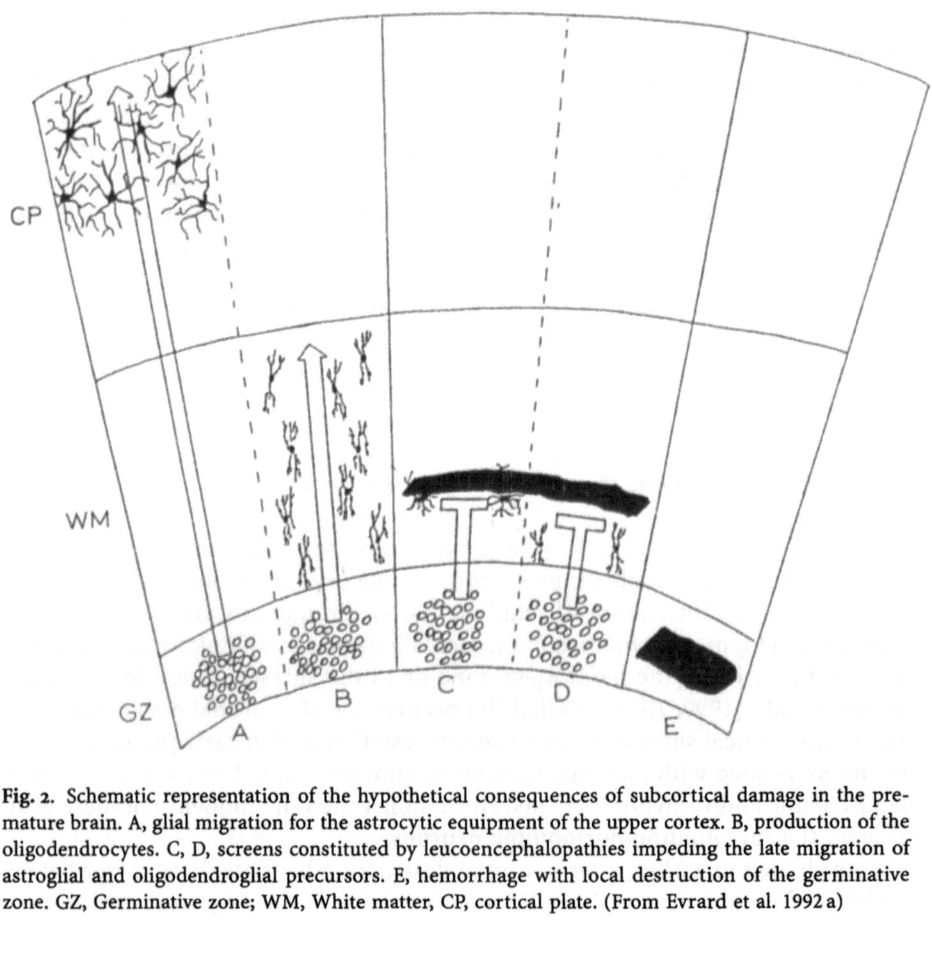

Fig. 2. Schematic representation of the hypothetical consequences of subcortical damage in the premature brain. A, glial migration for the astrocytic equipment of the upper cortex. B, production of the oligodendrocytes. C, D, screens constituted by leucoencephalopathies impeding the late migration of astroglial and oligodendroglial precursors. E, hemorrhage with local destruction of the germinative zone. GZ, Germinative zone; WM, White matter, CP, cortical plate. (From Evrard et al. 1992 a)

rhage and *periventricular leukomalacia* in the premature newborn (Fig. 2; Evrard et al. 1992 a, b, 1994). Germinal matrix hemorrhage destroys glial precursor cells in the germinative zone. The consequences are two-fold. Firstly, oligodendroglial progenitors are eliminated and/or their migration is disrupted, which may potentially result in later impaired myelination. Secondly, astrocytic precursors ultimately destined for the upper layers of neocortex may be destroyed and/or their migration disrupted, which may, thereby, impair cortical neuronal development. Major *hemorrhagic infarction in periventricular white matter*, as occurs with large, asymmetric intraventricular hemorrhage, and *periventricular leukomalacia* clearly have major deleterious impacts by interrupting projection and associative fibers. Additionally, oligodendroglia *in situ*, in preparation for myelination, may be destroyed, with subsequent impairment of myelination. Finally, these major hemorrhagic and nonhemorrhgaic lesions in the white matter may involve the neurons of the subplate which play important roles in cortical neuronal organization and in the development of cortical associative and projection connections. The importance of these mechanisms of late gliogenesis is highlighted when we consider that an appropriate astroglial contingent is a condition for the survival of a part of the neurons of the upper half of the cortex. Gressens et al. (1992 d) reported dramatic neuronal depletion (about 40 %) in the upper half of the mouse neocortex when, after completing normal neuronogenesis, the mitotic activity of the germinative zone is inhibited during the late developmental period, during which astroglial precursors of the upper cortical plate are produced in the germinative zone. These data also suggest a prolonged importance of Rakic's *protomap* for the germinative zone. These data suggest that the neuronal protomap is followed by an astroglial protomap that may have a crucial role for the cytoarchitectonic integrity of the upper half of the cortex. *Leucoencephalopathies in premature infants* could act as obstacles in the pathway of late migrating glia. These mechanisms could explain several neuropsychological consequences of prematurity.

The Excitotoxic Cascade at the Consecutive Neurodevelopmental Steps

Glutamate is the main excitotoxin during development. Among the different types of ionotropic glutamate receptors (see Johnston 1993), the N-methyl-D-aspartate (NMDA) receptor mediates excitotoxic neuronal injury during development. Glutamate metabotropic receptors have been reported to stimulate phosphoinositide hydrolysis (Sladeczek et al. 1985; Seren et al. 1989; Palmer et al. 1990) and to potentiate the NMDA effects (McDonald and Johnston 1992). These properties of glutamate receptors prompted us to investigate in developing rodent brains the effects of IBO, a competitive glutamatergic analog extracted from the *muscarina amanita* mushroom. IBO is an agonist of the NMDA complex receptor and of the metabotropic receptor. This section summarizes our recent data about the deviant neopallial architectural patterns produced with IBO (Fig. 2; Marret et al. 1994 a, b, 1995 a, b, d, 1996; Gressens et al. 1991; Evrard 1994).

1. Added to our whole mouse embryo cultures described above, IBO is devoid of any detectable influence when applied to this whole embryo culture (E8 – som 7/E9 – som 21; Marret et al. 1995 a, see above). Whole mouse embryo cultures (Gressens et al. 1992 a, 1993) were submitted to 10^{-7} to 10^{-3}M IBO at the 7-somite stage and cultured for 24 hours in a mixture of human and rat sera from the 7-somite stage to the 20–21-somite stages. During this developmental period of neural tube closure and early premigratory differentiation of the neuroepithelium, no macroscopic or light-microscopic effect of ibotenate has been detected (Marret et al. 1995 a, see above).

2. During maturation of neuronal layer V and during migration of neurons destined to granular and supragranular layers, newborn hamsters intracerebrally injected at P0 with 10 μg IBO display arrests of migrating neurons at different distances from the germinative zones. The resulting cytoarchitectonic patterns include periventricular nodular heterotopias, subcortical band heterotopias, and intracortical halts of migrating neurons (Marret et al. 1995 a, 1996). These migratory disturbances are isolated or associated with microgyria. The radial glial fibers and the extracellular matrix are exempt from any detectable damage, which could suggest a defect of their guiding role. The periventricular heterotopias induced by IBO are similar to the neuronal ectopic nodules encountered in Aicardi's and other human syndromes (for a review, see Evrard et al. 1992 b; Rorke 1994). The subcortical band heterotopias mimic some aspects of the human double cortex and lissencephalies (Dobyns et al. 1992).

3. After completion of migration of neurons destined for the infragranular layers, i.e., during the full settlement of supragranular layers, P0 mice injected with 5 μg IBO disclose a laminar neuronal depopulation of layer V–VIa sharply mimicking human microgyria (Marret et al. 1994 a, b, 1995 b, d). The "basic" lesion in most postmigratory polymicrogyrias is a postmigratory laminar depopulation predominating in the infragranular part of the cortical plate and especially in neocortical layers V–VIa (Barkovich et al. 1992; Dvorak and Feit 1977; Evrard et al. 1992 a, 1994; Gressens et al. 1991). Clinicopathologic correlations, including sequential ultrasound, from our European Multicenter Study confirm that perfusion failures leading to layered polymicrogyrias often happen after the end of neuronal migration and before the peak of gyration (between 20 and 28 weeks of gestation; Evrard 1988). Some cases of microgyria originate earlier, before the end of neuronal migration, and display late migratory arrests and laminar necrosis or depopulation. These variants of microgyria have been reviewed by Evrard et al. (1984).

4. Injected after completion of migration (P5–P10 in mouse), at the time when all neocortical neurons became aerobic, IBO produces a severe neuronal loss in all neocortical layers II, III, IV, V and VI (Fig. 3; Marret et al. 1994 a, b, 1995 a, b, d). This developmental lesional pattern mimics remarkably well the whole spectrum of neocortical lesions occurring in the at-term human newborn and fetuses: focal neocortical necroses throughout the whole cortical thickness, ulegyrias, multicystic encephalopathies, and porencephalic cysts. Furthermore, at this developmental stage, IBO induces the formation of white matter cysts,

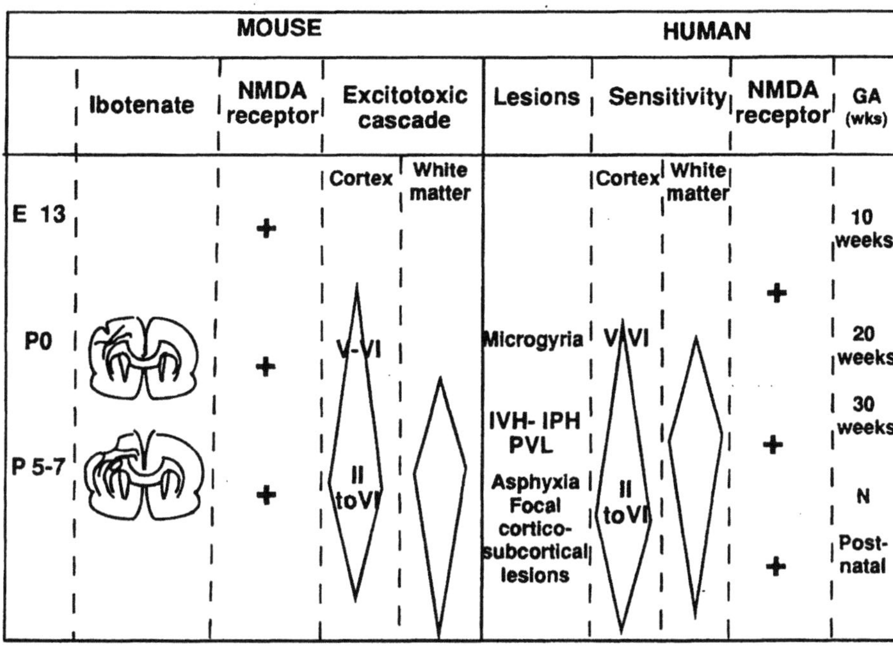

Fig. 3. Dysporphogenetic neocortical patterns induced by the excitotoxic cascade. E13, P0, P5–7, embryonic and postnatal ages in the mouse; GA, gestational ages in the human embryo and fetus (in weeks); N, at term in the human; +, presence of NMDA receptor; II to VI, numerals of the neocortical layers; IVH, intraventricular hemorrhage; IPH, periventricular hemorrhage; PVL, periventricular leucomalacia. The diamonds depict the intensity of the lesions described in the table and in the text. The schematic drawings under the title "ibotenate" are described in the text. This figure derives, after modifications, from a figure published in Marret et al. 1995 a

mimicing some aspects of the periventricular leukomalacias that are frequent in the premature newborn. In this mouse model, we observe a progressive regional decrease of white matter thickness with ventricular dilatation, following the reduction of cysts. This evolution in the classical evolution of cystic lesions of white matter in premature babies. We wish to stress again that maturation is species related: P5 in the mouse corresponds approximately to 24–30 weeks of gestation in humans as regards white matter astrocyte precursors, while mid-gestation in the human corresponds to birth in mouse as regards the completion of neuronal migration. All of these lesions are antagonized by DL-2-amino-7-phosphonoheptanoic acid, a competitive NMDA receptor antagonist, but not by L(+)-2-amino-3-phosphonopropionic acid, a metabotropic glutamate receptor (Marret et al. 1995 d). This study emphazises the dramatic developmental role of glutamate through NMDA receptor in the occurrence of neuronal migration disorders and gray and white matter damage observed in the cortex.

Preventions by Magnesium of Excitotoxic Neuronal Death in the Developing Brain

We also used intracerebral injections of IBO to analyze and quantify in the mouse at successive neurodevelopmental steps the protective effect of magnesium sulfate ($MgSO_4$), a non-competitive antagonist of the NMDA ionophore-complex receptor (Evrard 1994; Marret et al. 1994 b, 1995 a). Microgyria, the developmental neocortical pattern induced by IBO when injected at P0 (immediately after the postmigratory settlement of layer V), is not preventable by $MgSO_4$. Ulegyrias, porencephalic cysts, and other cortical and cortical-subcortical hypoxic-like lesions occuring after completion of the neocortical cytoarchitectonic development at P5 are remarkably preventable by a $MgSO_4$ systemic bolus before and, equally well, during a short period after the occurrence of excitotoxic aggression. Magnesium efficacy, at a peak at a P5 window, is diminishing from P5 to P10. The dose-effect study at P5 demonstrates optimal brain protection in 80 % of animals with 600 mg/kg, with no mortality due to $MgSO_4$. Lower doses provide a lower rate of protection; doses higher than 600 mg/kg provoke an increasing death culminating at 100 % mortality when 1,500 mg/kg of $MgSO_4$ is injected. The $MgSO_4$ protective effect appears after the developmental acquisition of two properties of the excitotoxic cascade, namely the coupling of the massive calcium influx with NMDA-receptor overstimulation and the predominance of magnesium-obliterable calcium channels.

This animal model supports the clinical intervention studies with magnesium in hypoxias/perfusion failures and is a basis for several suggestions for the design of these clinical studies (Nelson and Grether 1995). If maturation of the excitotoxic cascade follows the same sequence in the human, $MgSO_4$ protection of excitotoxicity is probably rather low before 28 weeks of gestational age and is probably low also at term, with a similar time window as in the animal model. The present study also suggests that excellent antiexcitotoxic effect in the human newborn could be abtained even if $MgSO_4$ is injected after the event. This model can be used to test other scavengers and glutamate antagonists that are candidates for protection of the developing human brain.

In conclusion, the distinct patterns induced by IBO in our study are strongly related to the stages of cerebral development and can be explained by changes in the subunits composition of the NMDA receptor during development, and its subsequent function. This IBO model provides tools to differentiate hypoxic from non-hypoxic excitotoxic lesions in fetuses and neonates and to investigate protective substances.

Acknowledgments

Most original data produced in our research laboratory and reviewed in this chapter have been produced with support from INSERM, France; Fondation de France; FNRS and FRSM, Belgium; NIH (Fogarty, NICHD and NINDS); USPHS;

Fondation pour la Recherche Médicale, and Fondation Madame Albert Froelich, Brussels, Belgium. We also wish to express our gratitude to the IPSEN Foundation and to the Nobel Symposia organization, Stockholm, Sweden, which provided us excellent opportunities to review this material. Acta Paediatrica Scandinavica (anniversary of the Pediatric Department, Karolinska Institute, Stockholm, Sweden) and the German Pediatric Neurology Association permitted us to use in the present chapter some material presented earlier to their audiences at their invitation.

References

Austin CP, Cepko CL (1990) Migration patterns in the developing mouse cortex. Development 89:8879-8883

Barkovich AJ, Gressens P, Evrard P (1992) Formation, maturation, and disorders of brain neocortex. Am J Neurorad 13:423-446

Carmeliet P, Mackman N, Moons L, Luther T, Gressens P, Van Vlaenderen I, Demunck H, Brieir G, Evrard P, Müller M, Risau W, Edgington T, Collen D (1996) Role of the cellular receptor tissue factor in embryonic blood vessel development. Nature 383:73-75

Caviness VS, Pinto-Lord MC, Evrard P (1981) The development of laminated pattern in the mammalian neocortex. In: Connelly PG, Brinkley LL, Carlson BM (eds) Morphogenesis and pattern formation. Raven Press, New York, 103-126

Dobyns WB, Elias ER, Newlin AC, Pagon MD, Ledbetter DH (1992) Causal heterogeneity in isolated lissencephaly. Neurology 42:1375-1388

Dvorak K, Feit J (1977) Migration of neuroblasts through partial necrosis of the cerebral cortex in newborn rats. Contribution to the problems of morphological development and developmental period of cerebral microgyria. Acta Neuropathol 38:203-212

Evrard P (1988) Les troubles du développement prénatal du cortex cérébral humain. Bulletin et Mémoires de l'Académie Royale de Médecine de Belgique 143:356-368

Evrard P (1994 a) ICNA/CNS Joint Congress. Presidential Symposium, San Francisco

Evrard P (1994 b) Ronald Mac Keith Lecture. British Pediatric Neurology Association, Birmingham

Evrard P, Minkowski A (eds) (1989) Developmental neurobiology. Raven Press, New York

Evrard P, Lyon G, Gadisseux JF (1984) Le développement prénatal du systéme nerveux et ses perturbations. Progr Néonatol 4:63-106

Evrard P, Kadhim HJ, Gadisseux JF (1989) Pathology of prenatal encephalopathies. In: French J (ed) Child neurology and developmental disabilities ICNA, Paul H. Brookes, Publisher, Baltimore, 153-176

Evrard P, Gressens P, Volpe JJ (1992 a) New concepts to understand the neurological consequences of subcortical lesions in premature brain. Biol Neonate 61:1-3

Evrard P, Miladi N, Bonnier C, Gressens P (1992 b) Normal and abnormal brain development. In: Rapin I, Segalowitz SJ (eds) Handbook of neuropsychology. Elsevier, Amsterdam, 11-44

Evrard P, Gressens P Kadhim HJ, Bonnier C, Gadisseux JF (1993) Environnement et développement du cerveau. Assoc Pédiat Langue Franç 77-86

Evrard P, Nassogne MC, Mukendi R, Bonnier C, Kadhim HJ, Gressens P, Gadisseux JF (1994) Environmental determinants of prenatal nervous system development, with particular emphasis on hypoxic-ischemic pathology. In: Lou HD, Greisen G, Larsen JL (eds) Brain lesions in the newborn. Royal Danish Academy of Sciences and Letters, Munksgaard, Copenhagen, 92-104

Evrard P, Marret S, Gressens P (1995) Envrionmental determinants of brain development, with a special emphasis on the excitotoxic cascade at the consecutive developmental steps. In: Rating D (ed) Aktuelle Neuropädiatrie 1994 Ciba-Geigy Verlag, Wehr, 22-33

Evrard P, Gadisseux JF, Gressens P (1996) Hypoxia opportunism during brain development. In: Wright LL, Merenstein GB, Hirtz D (eds) Acute Perinatal Asphyxia in Term Infants. National Institute of Child Health and Human Development, NIH Publication No. 96-3823, US Department of Health and Human Services 43-50, 55-60

Evrard P, Marret S, Gressens P (1997) Environmental and genetic determinants of neural migration and postmigratory survival. Acta Ped Scand Special Issue (Nobel Symposium "Nature - Nurture"), in press

Gadisseux JF, Evrard P (1985) Glial-neuronal relationship in the developing central nervous system. A histochemical-electron microscope study of radial glial cell particulate glycogen in normal and reeler mice and the human fetus. Dev Neurosci 7:12-32

Gadisseux JF, Kadhim HJ, van den Bosch P, Caviness VS, Evrard P (1990) Neuron migration within the radial fiber system of the developing murine cerebrum. Dev Brain Res 52:39-56

Gofflot F, Nassogne MC, Etzion T, Van-Maele-Fabry G, Evrard P, Picard JJ (1995) In vitro neuroteratogenicity of valproic acid (VPA) and 4-EN-VPA. Neurotoxicol Teratol 17:425-435

Gressens P, Evrard P (1993) The glial fascicle: an ontogenic and phylogenic unit guiding, supplying and distributing mammalian cortical neurons. Dev Brain Res 76:272-277

Gressens P, Cilio MR, Schlögel X, Evrard P (1991) Les mécanismes de l'hypoxie pendant la vie foetale et néonatale. Progr Néonatol 11:203-226

Gressens P, Gofflot F, Van Maele-Fabry G, Misson JP, Gadisseux JF, Picard J, Evrard P (1992 a) Early neurogenesis and neural teratogenesis on whole mouse embryo cultures: Histochemical, immunocytological and electron microscopic study of the premigratory neuronal-glial units in normal mouse embryo and under the influence of retinoic acid. J Neuropathol Exp Neurol 51:206-219

Gressens P, Kosofsky BE, Evrard P (1992 b) Cocaine-induced disturbances of corticogenesis in the developing murine brain. Neurosci Lett 140:113-116

Gressens P, Lammens M, Picard JJ, Evrard P (1992 c) Ethanol-induced disturbances of gliogenesis and neuronogenesis in the developing murine brain. Alcohol Alcohol 27:219-226

Gressens P, Richelme C, Kadhim H, Gadisseux JF, Evrard P (1992 d) The germinative zone produces most cortical astrocytes after neuronal migration in developing mammalian brain. Biol Neonate 61:4-24

Gressens P, Hill JM, Gozes I, Fridkin M, Brenneman DE (1993) Growth factor function of vasoactive intestinal peptide in whole cultured mouse embryos. Nature 362:155-158

Gressens P, Hill JM, Paindaveine B, Gozes I, Fridkin M, Brenneman DE (1994) Blockade of vasoactive intestinal peptide function in early mouse embryo induces a severe microcephaly. J Clin Invest 94:2020-2027

Gressens P, Marret S, Evrard P (1996) Developmental spectrum of the excitotoxic cascade induced by ibotenate: a model of hypoxic insults in fetuses and neonates. Neuropathol Appl Neurol 22:498-502

Gressens P, Marret P, Evrard P (1997 a) La neuroprotection du cerveau du nouveau-né. In: Relier JP (ed) Progrès néonatologie. Volume 17, Karger, Basel, in press

Gressens P, Paindaveine B, Hill JM, Brenneman DE, Evrard P (1997 b) Growth factor properties of VIP during early brain development; whole embryo cultures and in vivo sudies. Ann NY Acad Sci, in press

Hill JM, McCune SK, Alvero RJ, Glazner GW, Henins KA, Stanziale SF, Keimovitz JR, Brenneman ADE (1996) Maternal vasoactive intestinal peptide and the regulation of embryonic growth in the rodent. J Clin Invest 97:1-7

Jammes JL, Gilles FH (1983) Telencephalic development: matrix volume and isocortex and allocortex surface area. In: Gilles FH, Leviton A, Dooling EC (eds) The developing human brain. Growth and epidemiologic neuropathology. Boston, J Wright PSG Inc., 87-93

Johnston MV (1993) Cellular alterations associated with perinatal asphyxia. Clin Invest Med 62:122-132

Kosofsky BE, Wilkins AS, Gressens P, Evrard P (1994) Transplacental cocaine exposure: a mouse model demonstrating neuroanatomic and behavioral abnormalities. J Child Neurol 9:234-241

Kuban KC, Gilles FH (1985) Human telencephalic angiogenesis. Ann Neurol 17:539-548

Marin-Padilla MJ (1990) The pyramidal cell and its local-circuit interneurons: a hypothetical unit of the mammalian neocortex. Cog Neurosci 38:183-190

Marin-Padilla M (1993) Pathogenesis of later-acquired leptomeningeal heterotopias and secondary cortical alterations. In: Galaburda AM (ed) Dyslexia and development. Neurobiological aspects of extraordinary brain. Cambridge, MA, Harvard University Press

Marret S, Mukendi R, Gadisseux JF, Gressens P, Evrard P (1994 a) Effect of ibotenate on brain development: an excitotoxic mouse model of microgyria and posthypoxic like lesions. Ped Neurol 11:104

Marret S, Mukendi R, Gadisseux JF, Gressens P, Evrard P (1994 b) Magnesium protects neonatal mouse brain from posthypoxic like damages induced by ibotenate in mouse pups. Ann Neurol 36:99

Marret S, Gadisseux JF, Gressens P, Evrard P (1995 a) Prevention by magnesium of excitotoxic neuronal death in developing brain. An animal model for the clinical intervention studies. Dev Med Child Neurol 37:473–484

Marret S, Gressens P, Evrard P (1995 b) Developmental spectrum of the excitotoxic cascade induced by ibotenate: a model of hypoxic insult in fetuses and neonates. Dev Med Child Neurol 37:81

Marret S, Gressens P, Van-Maele G, Evrard P (1995 c) Effects of caffeine on the maturation of the early neural germinative epithelium in mouse whole embryo cultures. Dev Med Child Neurol 37:82

Marret S, Mukendi R, Gadisseux JF, Gressens P, Evrard P (1995 d) Effect of ibotenate on brain development: an excitotoxic mouse model of microgyria and posthypoxic like lesions. J Neuropathol Exp Neurol 54:358–370

Marret S, Gressens P, Evrard P (1996) Neuronal migration disorders induced by ibotenate in the neocortex. Proc Natl Acad Sci USA 93:15463–15468

McDonald JW, Johnston MV (1992) Neuroprotective synergism of 2-amino-3-phosphonopropionate (D, L-AP3) and MK-801 against ibotenate induced brain injury. Neurosci Lett 145:213–216

Mountcastle VB (1957) Modality and topographic properties of single neurons of cat's somatic sensory cortex. J Neurophysiol 20:408–434

Nassogne MC, Evrard P, Courtoy P (1996) Selective neuronal toxicity of cocaine in embryonic mouse brain cocultures. Proc Natl Acad Sci USA 92:11029–11033

Nelson KB, Grether JK (1995) Can magnesium sulfate reduce the risk of cerebral palsy in very low birthweight infants? Pediatrics 95:263–269

Norman M, O'Kusky JR (1986) The growth and development of microvasculature in human cerebral cortex. J Neuropathol Exp Neurol 45:222–232

Palmer E, Nangel-Taylor K, Krause JD, Roxas A, Cotman CW (1990) Changes in excitatory amino acid modulation of phosphoinositide metabolism during development. Dev Brain Res 51:132–134

Pellerin L, Magistretti P (1994) Glutamate uptake into astrocytes stimulates aerobic glycolysis: a mechanism coupling neuronal activity to glucose utilization. Proc Natl Acad Sci USA 91:10625–10629

Rakic P (1971) Guidance of neurons migrating to the fetal monkey neocortex. Brain Res 33:471–476

Richman DP, Stewart RM, Caviness VS (1974) Cell cycle parameters and patterns of nuclear movement in the neocortical proliferative zone of the fetal mouse. J Neuropathol Exp Neurol 33:374–384

Rorke LB (1994) The role of disordered genetic control of neurogenesis in the pathogenesis of migration disorders. J Neuropathol Exp Neurol 53:105–117

Seren MS, Aldino C, Zanoni R, Leon A, Nicoletti F (1989) Stimulation of inositol phospholipid hydrolysis by escitatory amino acids is enhanced in brain slices from vulnerable regions after transient global ischemia. J Neurochem 53:1700–1705

Sladeczek F, Pin J, Recasens M, Bockaert J, Weiss S (1985) Glutamate stimulates inositol phosphate formation in striatal neurons. Nature 317:717–719

Takahashi T, Nowakowski RS, Caviness VS (1993) Cell cycle parameters and patterns of nuclear movement in the neocortical proliferative zone of the fetal mouse. J Neurosci 13:820–833

Tan SS, Breen S (1993) Radial mosaicism and tangential cell dispersion both contribute to mouse cortical development. Nature 362:638–640

Walsch C, Cepko CL (1992) Widespread dispersion of neuronal clones across functional regions of the cerebral cortex. Science 255:434–440

Yuste R, Peinado A, Katz LC (1992) Neuronal domains in developing neocortex. Science 257:665–669

Regulators of Neuronal Survival (Bcl-2, Bax, c-Jun) in Prenatal and Postnatal Human Frontal and Temporal Lobes in Normal and Down Syndrome Brain

K.E. Wisniewski, E. Kida, I. Kuchna, T. Wierzba-Bobrowicz, and M. Dambska

Abstract

Although the molecular basis of programmed cell death, a pivotal process for the proper development, organization and future functioning of the brain, has been studied extensively in different experimental models and in various species, it is not entirely understood. The principles that govern this process in the developing human brain are even less elucidated. To provide more insight into the mechanisms underlying programmed cell death in the human brain, we have studied the temporal and spatial pattern of immunolocalization in the developing normal and Down syndrome brain of two proteins that are supposed to act either as a neuronal survival promoter-Bcl-2 or a neuronal survival suppressor-Bax. Our data show a dramatic decrease in immunoreactivity to Bcl-2 and, at the same time, a gradual increase in immunoreactivity to Bax in the human brain around 30–32 week's gestation; according to recent data, this gestational period is accompanied by the most intense reduction of cortical neurons. However, our obser vations also suggest that at least at certain stages of human brain development, both proteins, and especially Bax, may also be involved in the final differentiation and maturation of central nervous system neurons. In addition, altered patterns of Bcl-2 and, to a much lesser degree, also of Bax immunoreactivity in developing Down syndrome brain allow us to propose that mechanisms underlying programmed cell death may contribute to the pathology in Down syndrome brain.

The Position of Bcl-2 and Bax in Programmed Cell Death

Programmed cell death (PCD) mediated by apoptosis is a genetically controlled process immanently associated with the normal development of multicellular organisms. By means of PCD, cells that are generated in excess or that are needed only transiently are eliminated, which allows for the proper organization and functioning of many tissues and organs, including the central nervous system (CNS; Wyllie et al. 1980; Arends and Wyllie 1991). Apoptosis is also important for maintaining homeostasis in adult organisms. It is believed that apoptosis represents a favorable type of cell death, as it maintains the membranes' integrity and is not followed by an inflammatory response. However, whether apoptosis repre-

A. M. Galaburda / Y. Christen (Eds.)
Normal and Abnormal Development of Cortex
© Springer-Verlag Berlin Heidelberg New York 1997

sents a unique mechanism of cell death during development is still a matter of debate (Schwartz et al. 1993; Yaginuma et al. 1996).

The morphological events underlying the apoptotic process are well documented and are similar in most mammalian cells (Kerr et al. 1972; Wyllie et al. 1980). Structural alterations include chromatin condensation and margination, cell shrinkage, membrane blebbing and, finally, the formation of membrane-bound apototic bodies, which are removed by neighboring cells or macrophages. However, the molecular basis of this process is still far from being understood. In contrast to necrosis, apoptosis is an active process and requires the expression of specific, apoptosis-linked genes (Ellis et al. 1991). Ced-3, a gene identified in a nematode, *Caenorhabditis elegans*, is one of the best characterized genes essential for PCD (Ellis and Horvitz 1986). The product of this gene shows a homology with mammalian cysteine protease, interleukin-1β-converting enzyme (ICE; Yuan et al. 1993), which cleaves pro-interleukin-1β to a biologically active form (Thornberry et al. 1992). Thus, it has been suggested that the activation of specific proteases, members of the CED-3/ICE family, may be implicated in PCD (Tewari and Dixit 1995). Currently identified members of the ICE/CED-3 protease family show structural homology either to ICE (ICE-rel-II, ICE-rel-III) or to CED-3 (ICH-1/Nedd2, CPP32/apopain, Mch2, Mch3). Human cell death genes CPP32, Mch2, Mch3, and Ich-1 have been mapped recently (Tiso et al. 1996), and their expression has been described in both embryonic and adult human cell lines (Wang et al. 1994; Fernandes-Alnemri et al. 1994). Recently, two other apoptosis-linked genes have been identified. ALG-2, coding for a Ca^{2+}-binding protein, was found to be required for T-cell receptor-, Fas-, and glucocorticoid-induced cell death. ALG-3, homologous to the familial Alzheimer's disease (AD) gene, STM2, showed an opposite, protective effect against apoptotic events induced by these stimuli (Vito et al. 1996). Thus, it is becoming evident that numerous complex mechanisms, including proteolytic processes and Ca^{2+}-regulating signals, are involved in PCD.

Apart from "death genes," whose expression is necessary for PCD to occur, a subset of genes has been identified whose products protect the cell against death in response to various apoptotic stimuli. The counterpart in mammals of the ced-9 gene *Caenorhabditis elegans*, which prevents the death of all cells genetically determined to undergo PCD, is bcl-2 (Hengartner and Horvitz 1994). The bcl-2 (B-cell leukemia/lymphoma) gene was first identified at the chromosomal breakpoint of the t(14;18) chromosomal translocation in follicular B-cell lymphoma (Tsujimoto et al. 1985; Bakhshi et al. 1985; Cleary and Sklar 1985). The product of this gene, Bcl-2, is a 25-kDa integral membrane protein, localized in the cell to the nuclear envelope, endoplasmic reticulum and outer mitochondrial membrane (Monagham et al. 1992; Akao et al. 1994). This protein was found to be able to suppress cell death following a variety of stimuli, but not to prevent negative selection in thymocytes or cell death induced by IL-2 or IL-6 depletion (Nunez et al. 1990; Sentman et al. 1991). The mechanisms of Bcl-2 protection against various apoptotic stimuli are not clear. Bcl-2 might mediate its protective action via an antioxidant pathway (Hockenbery et al. 1993), either acting as a pro-oxidant and influencing levels of reactive oxygen intermediates (Steinman 1995) or regulating Ca^{2+} fluxes and interfering with Ca^{2+} signaling (Baffy et al. 1993; Lam et al. 1994).

Previously performed immunocytochemical studies revealed a specific and organized pattern of Bcl-2 localization in various organs of both adult and embryonic tissues, suggesting participation of this protein in cell development, maturation and terminal differentiation (Hockenbery et al. 1991; Merry et al. 1994; Veis Novack and Korsmeyer 1994).

Recently, other genes regulating apoptosis and belonging to the growing bcl-2 family have been identified (Korsmeyer 1995). The products of these genes show homology with Bcl-2, mainly within two highly conserved regions: the BH1 and BH2 domains. However, their function in the process of apoptosis varies. Some of these proteins are able to repress apoptosis (Bcl-x_L,); meanwhile the others (Bax, Bcl-x_S, Bad, Bak) have been found to promote cell death (Oltvai et al. 1993; Gonzalez-Garcia et al. 1994; Yang et al. 1995; Chittenden et al. 1995; Minn et al. 1996). Like Bcl-2, Bax (Bcl-2-associated X protein) is an integral membrane protein, with an approximate Mw of 21 kDa. Bax mRNA and immunoreactivity to Bax were disclosed in various murine tissues and organs (Oltvai et al. 1993; Krajewski et al. 1994). Bax forms heterodimers with Bcl-2 and homodimers with Bax. The ratio of Bcl-2 to Bax might determine the further fate of the cell after PCD signal: either cell survival in excess of Bcl-2 or cell death in excess of Bax (Oltvai et al. 1993; Oltvai and Korsmeyer 1994).

Some data have suggested that interactions between the proteins within the Bcl-2 family, and particularly their dimerization, substantially determine their function in PCD. Site-specific mutations of their genes alter or abrogate their properties (Oltvai et al. 1993; Yin et al. 1994), affect protein trafficking and determine cell-specific patterns of response to apoptotic stimuli (Zhu et al. 1996). However, it has also been suggested that, at least in certain cell types, binding of Bax by Bcl-2 is not sufficient to prevent apoptosis (Zhu et al. 1996). Recently, it was found that the N-terminal region of Bcl-2, encompassing five amino acid residues, may be critical to mediate protein-protein interactions. This motif is shared by other members of the Bcl-2-family but not by those that potentiate apoptosis (Lee et al. 1996).

Bcl-2 and Bax During Normal Development of the Frontal and Temporal Cortex in Humans

Around 50–70 % of neurons die by apoptosis during normal development of the CNS in mammals (Cowan et al. 1984; Blaschke et al. 1996). The death or survival of the neuron is determined mostly by competition for target-derived neurotrophic factors. The involvement of Bcl-2 in neuronal survival after a PCD signal was demonstrated in numerous experiments. Bcl-2 was able to protect sympathetic and sensory neurons against nerve growth factor, brain-derived neurotrophic factor and neurotrophin-3 withdrawals, but could not rescue ciliary neurons from death after ciliary neurotrophic factor withdrawal (Garcia et al. 1992; Allsopp et al. 1993). Bcl-2 also inhibited the death of the central neural cell line caused by serum and growth factor withdrawal, the calcium ionophore A23187, glucose withdrawal, membrane peroxidation and, to some extent, free radical-induced damage (Zhong et al. 1993).

Evidence for the role of Bcl-2 in the process of normal brain development, organization and maturation was provided by studies in transgenic mice that either overexpressed or were deficient in Bcl-2. Overexpression of Bcl-2 reduced the normal neuronal loss during brain development and led to hypertrophy of nerves in experimental animals (Dubois-Dauphin et al. 1994; Martinou et al. 1994). According to an earlier report, bcl-2 -/- mice showed growth retardation, early mortality, massive cell death in lymphoid organs, and polycystic kidney and hair hypopigmentation, but no significant CNS alterations (Veis et al. 1993). However, degeneration of motor, sensory, and sympathetic neurons was recently documented in another model of bcl-2-deficient mice (Michaelidis et al. 1996). Significantly, prenatal neuronal survival was not altered. CNS abnormalities occured in these animals early in the postnatal period, thus after the physiological cell death period.

In developing murine brain, Bcl-2 was found to be widely expressed in the prenatal period, whereas it was confined mostly to the granule cells of the cerebellum, dentate gyrus of the hippocampus and olfactory bulb in the postnatal period (Merry et al. 1994). Bcl-2 was also observed in certain neuronal populations of developing murine nervous system beyond the period of PCD, which suggests that this protein might be involved in biological processes other than PCD. Except for our preliminary report (Wisniewski and Kida 1994), Bcl-2 in human fetal brain has not been immunolocalized. Our current data indicate that the pattern of Bcl-2 immunolocalization in human fetal brain shows substantial differences from that documented earlier in mice brain. From the earliest periods of human fetal brain development studied, 10 weeks of gestation (wg), Bcl-2-immunoreactive cells were present in the ventricular zone, intermediate zone and cortical plate. However, although the level of Bcl-2 remained relatively constant in the ventricular zone until late in gestation, the immunoreactivity to Bcl-2 was reduced early in this part of the germinal layer that formed the ganglionic eminence. Moreover, in contrast to homogeneous, widespread immunostaining of the ventricular zone, the ganglionic eminence demonstrated small clusters of Bcl-2-positive cells often separated by unstained cells. Immunoreactivity in the ganglionic eminence persisted for longer periods, mainly in its lower part. A similar reduction of immunoreactivity to Bcl-2 was observed in the intermediate zone, from around 18–22 wg. Numerous Bcl-2-positive cells were visible much longer in the subplate zone, the brain region implicated in the formation of the first subcortical connections and most probably important for cortical plasticity in the later periods of life (Kostović et al. 1989; McConnell et al. 1989). Although the subplate zone contains the first postmitotic neurons in the cerebral cortex, maturation of the cells in this layer is prolonged until even neonatal life (Luskin and Shatz 1985). In contrast to the subplate zone, only weak Bcl-2 immunostaining was found in the cells of the subpial granular layer and especially in Cajal-Retzius neurons, both present transiently in the brain during gestational life (Marin-Padilla 1984; Gadisseux et al. 1992). Although the immunoreactivity to Bcl-2 in the neocortex and archicortex was already visible at 10 wg, these cortical areas showed a dissimilar time course and pattern of Bcl-2 immunostaining in the later gestational periods. Immunostaining was the strongest in the neocortex, and more widespread than in the other brain areas examined. The immunoreactivity to Bcl-2 of the neocortex gradually increased, from its initial uniform distribution in the cells forming the cortical plate.

From around 12–14 wg, the cells located in the upper neocortical zone, just below the marginal zone, became the most intensely immunostained. From around 16–17 wg, a thin layer of strongly immunostained small cells with immunopositive dendritic processes appeared in the lower part of the middle cortical zone of the frontal cortex. Later on, at around 18–20 wg, the Bcl-2-immunopositive cells showed more organized, laminar distribution and a gradual reduction of immunostaining in the upper-middle, and increasing immunostaining in the middle and lower, cortical zones. This laminar organization of Bcl-2-immunopositive cells could be also detected in the temporal cortex, but in this brain region, the immunostaining of the upper and of the deepest cortical zones predominated, being recognizable from around 19 wg, thus later than in the frontal cortex. Around 32 wg the immunoreactivity to Bcl-2 in the neocortex dramatically fell. From this gestational period until postnatal life, it was mostly the IInd cortical layer that retained numerous strongly immunopositive cells. Distinctly weaker immunoreactivity to Bcl-2, confined to smaller neuronal population, was present in the IIIrd and V–VIth cortical layers.

This pattern of immunostaining was present in the neocortex until the second year of life; however, the number of immunostained neurons and the intensity of immunoreactivity gradually diminished (Fig. 1). Although the exact time course of apoptosis in the human fetal cortex is unknown, recently published data suggest that substantial overproduction and subsequent reduction of cortical neurons occur mainly during the last trimester of gestation, after the 32nd wg (Rabinowicz et al. 1996). Thus, the dramatic drop of Bcl-2 immunoreactivity found by us in the human fetal neocortex starting from around 30–32 wg shows a strong correlation with the increased rate of death of neocortical neurons observed during this gestational period. It is also of interest that the laminar distribution of Bcl-2 immunoreactivity in the neocortex was due to the immunostaining of small neurons lacking differentiation characteristics. Bcl-2 immunostaining was either absent or only marginal in more differentiated, large neurons that were visible in the neocortex from around 18–20 wg, as was evidenced by MAP-2 and Bax immunostaining. The laminar pattern of Bcl-2 immunostaining observed during midgestation in human fetuses did not correlate with the time course of neuronal arrival to the neocortex. At the same stage of gestation, prominent immunoreactivity was found in the future IInd cortical layer, that is, that neocortical layer bearing the neurons that arrive late during corticogenesis, and in the middle and deep neocortical zones, which contain neurons that arrive at

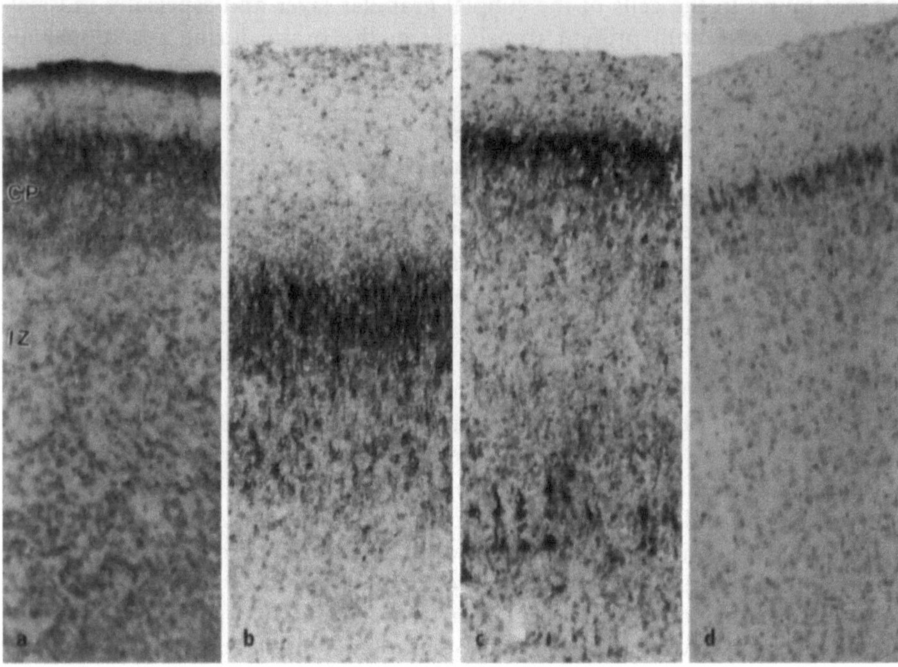

Fig. 1. Immunoreactivity to Bcl-2 of the human temporal neocortex at various periods of gestational life. a) 12 weeks of gestation (wg); b) 22 wg; c) 26–28 wg; d) 34 wg. x 200

the cortical plate earlier (Volpe 1977; Sara 1986). These observations suggest that young neocortical neurons may need protection against apoptotic stimuli at different, but very specific, periods of their life depending on their final destination. The prolonged immunoreactivity to Bcl-2 found in the IInd neocortical layer of human fetal brain could in this respect be attributed to the late development and final organization of this particular layer, which shows numerous immature neurons also in the postnatal period. Immunoreactivity to Bcl-2 in the hippocampal formation was also the most prominent in small, undifferentiated neurons. The strongest Bcl-2 immunostaining showed small neurons in the IInd layer of the entorhinal cortex around 26–28 wg; however, the immunoreactivity to Bcl-2 disappeared earlier in the hippocampal formation than in the neocortex and was not visible at birth, except for in the dentate gyrus. This observation is in agreement with the earlier accomplishment of final differentiation by the hippocampal structures than by the neocortex, except for the dentate gyrus, which undergoes final differentiation during postnatal life.

As was pointed out above, the immunolocalization of Bax, studied in numerous organs of adult mice, did not always parallel that of Bcl-2 and was not confined to tissues characterized by high apoptotic death rates (Krajewski et al. 1994). In adult murine cerebral cortex, Bax was often observed in large and smaller neurons (Krajewski et al. 1994). Our data extend these findings and pro-

Table 1. Schematic representation of immunoreactivity to Bcl-2 and Bax in various areas of the frontal and temporal lobes.

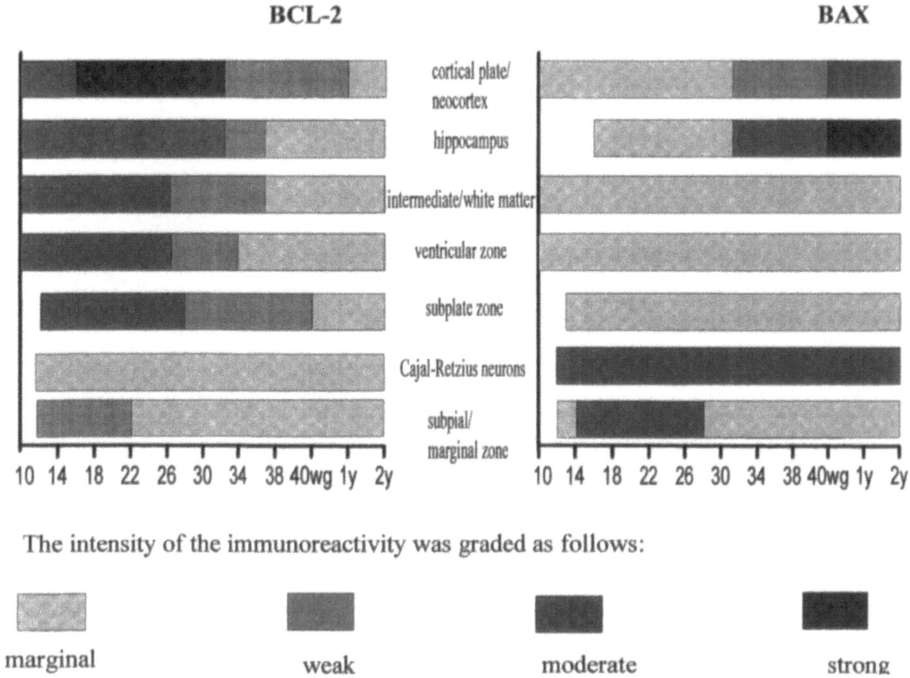

vide evidence that the pattern and time course of Bax immunostaining of the fetal and postnatal human cortex are in many instances strikingly dissimilar or even opposite to those of Bcl-2 (Table 1). Generally, except for a short early gestational period when Bax was weakly expressed mostly in the upper cortical layer, Bax was found in more mature, large neurons in many of the neocortical and archicortical regions studied (Fig. 2). Moreover, in contrast to immunoreactivity to Bcl-2, immunoreactivity to Bax gradually increased, being the most widespread in the second year of life. This relation of Bax immunoreactivity to the stage of neuronal development was especially evident in the hippocampus. Stronger immunoreactivity to Bax first appeared around 30–32 wg in the cytoplasm of differentiated neurons of the CA2 and CA3 sectors, neurons that differentiate earlier. Similarly strongly immunostained neurons appeared later, in the postnatal period, in those hippocampal structures whose differentiation and maturation is prolonged (CA1 sector and dentate gyrus). The immunoreactivity to Bax was, in contrast to immunoreactivity to Bcl-2, only weak in the ventricular, intermediate and subplate zone, except for some strongly stained large neurons in the subplate zone and white matter. However, much stronger immunoreactivity to Bax than to Bcl-2 revealed cells located in the subplate/marginal zone and Cajal-Retzius neurons, both of which are present transiently in the developing human brain. Thus, the constant presence of Bax in cells genetically predetermined to die suggests that at least for this neuronal population in the developing human

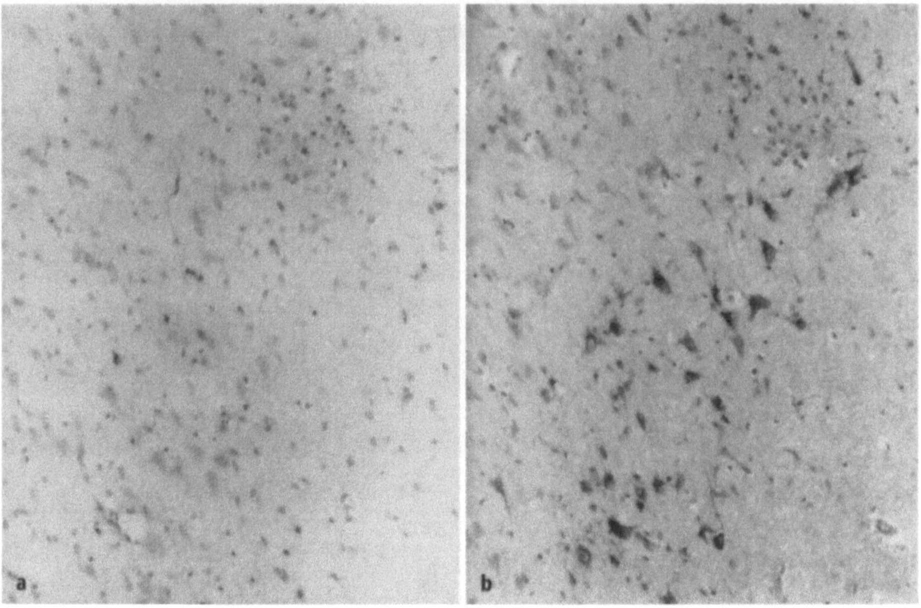

Fig. 2. Different pattern of immunoreactivity to Bcl-2 (a) and Bax (b) of neurons in the entorhinal cortex. Perinatal period. x 200

brain, Bax may play the role of a specific "death factor." Moreover, a gradual increase of Bax immunoreactivity started in the human fetal cortex from around 30–32 wg, paralleling the dramatic decrease in Bcl-2 immunoreactivity and at the same time the onset of a prominent apoptotic cell death.

These observations appear to support the role of Bax as a promoter of cell death in the developing human brain. However, it appears that a gradual increase in immunoreactivity to Bax during the postnatal period of human brain development, as well as the type and the topography of immunostained cells, cannot be regarded as a sign of apoptotic pathway induction. Furthermore, in the postnatal period, Bax was immunolocalized mostly to large, already differentiated neurons, as was evidenced by their morphological characteristics. Thus, it can be postulated tentatively that, especially during this particular period of human cortex development, Bax migth be implicated in the process of terminal differentiation and maturation of neurons rather than in their apoptotic death. It was found recently that, in bax-/- mice, there is a 51 % increase in the total number of facial motor neurons, most probably due to altered PCD in the absence of Bax (Deckwerth et al. 1996). However, Bax is not necessary for all cell deaths to occur, and not all cell lineages undergo hyperplasia in bax-/- mice, which suggests that the control of cell death may differ with lineage and stage of development (Deckwerth et al. 1996). It is of importance that the bax-/- mice studied to date showed no behavioral changes that might be related to the reduction of developmental neuronal death. Moreover, it was recently proposed that the effect of Bax on neuronal survival is different from its apoptosis-accelerating action, which was observed in IL-3-dependent cells studied previously. Bax was found not only to promote neuronal survival in vitro after withdrawal of neurotrophic factors but its protective action also dominated that of neurotrophic factors (Middleton et al. 1996). This intriguing possibility that Bax may also induce survival-promoting action for neurons needs further studies.

Morphological Alterations of Down Syndrome (DS) Brain Development

The incidence of DS is 1 to 1.4 per 1,000 live births. DS is caused either by full or partial trisomy of chromosome 21, or a mosaic of normal and trisomic cells (Kemper 1988). It is well documented that the DS genotype is associated with a variety of phenotypes, of which the most common are mental retardation and neonatal hypotonia. Morphological abnormalities in DS brain were reported for the first time at the beginning of this century (Apert 1914; Davidoff 1928). However, there is still no consensus as to the time at which morphological abnormalities start in DS brain, what the typical pattern of brain alterations is and the manner in which these changes contribute to mental retardation and early development of AD pathology, and their direct cause. The most prominent abnormalities are found relatively late during development of children with DS. One of the cardinal signs of DS, growth retardation, occurs between six months and three years of age (Sara et al. 1983; Cronk et al. 1978; Cronk et al. 1988). DS subjects older than two years of age also demonstrate a deficiency of insulin-like growth factor 1 receptors (Sara et al. 1983). It is also in postnatal life, mostly in the second part of midinfancy, that microcephaly and a 20 to 50 % reduction in brain weight develop in individuals with DS (Crome et al. 1996; Wisniewski 1990). Gross brain pathology, decreased neuronal densities, delayed myelination, dendritic alterations and synaptic pathology were also found to occur mostly in infancy and early childhood (Colon 1972; Marin-Padilla 1976; Suetsugu and Mehraein 1980; Takashima et al. 1981; Becker et al. 1986, 1991; Schmidt-Sidor et al. 1990; Wisniewski et al. 1984, 1986; Wisniewski 1990). No changes in the concentration of neural cell adhesion molecules, the total composition of gangliosides, the major classes of lipids, or the activity of choline acetyltransferase, and only a slightly higher proportion of phosphatidylcholine were found in fetal DS brains studied at 19.5 to 24.5 weeks of gestation (Brooksbank et al. 1989). However, the growth of numerous organs was already found to be significantly lower in second-trimester fetuses with DS than in controls (Fitzsimmons et al. 1990). There are data suggesting that development of the brain in a subject with DS is also already compromised during the gestational period. Anomalies in the polyunsaturated fatty acid composition of phosphoglycerides were detected at the early stage of DS brain development (Brooksbank et al. 1985). Abnormal electric membrane properties of fetal DS dorsal root ganglion neurons have been also reported (Scott et al. 1982). In addition, DS fetal brain tissues showed increased lipoperoxidation as well as elevation of the cytosolic Cu/Zn-dependent superoxide dismutase, but not followed by the commonly observed compensatory increase of glutathione peroxidase activity (Brooksbank and Balazs 1984). Moreover, ultrastructural studies revealed a diminished number and abnormal morphology of synapses in DS brain at birth (Wisniewski et al. 1984, 1986) and a less consistent pattern of synaptic development, with reduced synaptic parameters and increased number of primitive contacts starting between 32 and 34 weeks of gestation (Petit et al. 1984). All these data strongly support the notion that, although the abnormalities of DS brain become clearly evident in postnatal life, they originate much earlier, during fetal life. Our data showing discrete alteration of synaptophysin immunostaining in the second half of gestation in DS fetuses (Wisniewski and Kida 1994), the results of morphometric analysis of the development of the superior temporal gyrus in DS (Golden and Hyman 1994), as well as earlier data showing no gross alteration in DS brain at 15 to 22 weeks of gestation (Schmidt-Sidor et al. 1990) all suggest that the critical period for DS brain development may take place late in the second or in the last trimester of gestational life.

The Pattern of Bcl-2 and Bax Immunolocalization in the Frontal and Temporal Cortex of Developing Down Syndrome Brain

Evidence is growing that dysregulation of genes involved in PCD may not only affect normal neuronal development but also underlie neurodegenerative processes in the brain such as those encountered in ischemia, amyotrophic lateral sclerosis or AD type of pathology (Satou et al. 1995; Hara et al. 1996; Mu et al. 1996). A defect in the metabolism of reactive oxygen species leading to apoptosis has been described in cortical neurons of the fetal DS brain, which might contribute to both mental retardation and development of AD pathology in adult life in DS subjects (Brooksbank and Balazs 1984; Busciglio and Yankner 1995). Dysregulation of mechanisms underlying PCD in developing DS brain was also suggested by our earlier data (Wisniewski and Kida 1994).

In contrast to a relatively uniform pattern of Bcl-2 and Bax immunolocalization in developing normal human brain, a less consistent pattern and marked interindividual differences were found in DS brain. The most commonly observed and the most prominent difference in the distribution of Bcl-2 between the normal and DS brain pertained to the IInd cortical layer. In this neocortical layer, DS subjects showed Bcl-2 immunopositivity less often than controls, especially in the perinatal and postnatal periods (Fig. 3). It is interesting that this layer is distinctly affected in DS subjects and, apart from the internal granular layer, shows a marked neuronal loss in postnatal life (Galaburda and Kemper 1979; Ross et al. 1984; Wisniewski et al. 1986). This associative cortex contains classes of cortical interneurons that are important for intrinsic local circuitry and corticocortical connections, most probably of inhibitory character. Thus, it appears that reduced immunoreactivity to Bcl-2 within the external granular layer in DS subjects might contribute to further neuronal loss in this area, as a

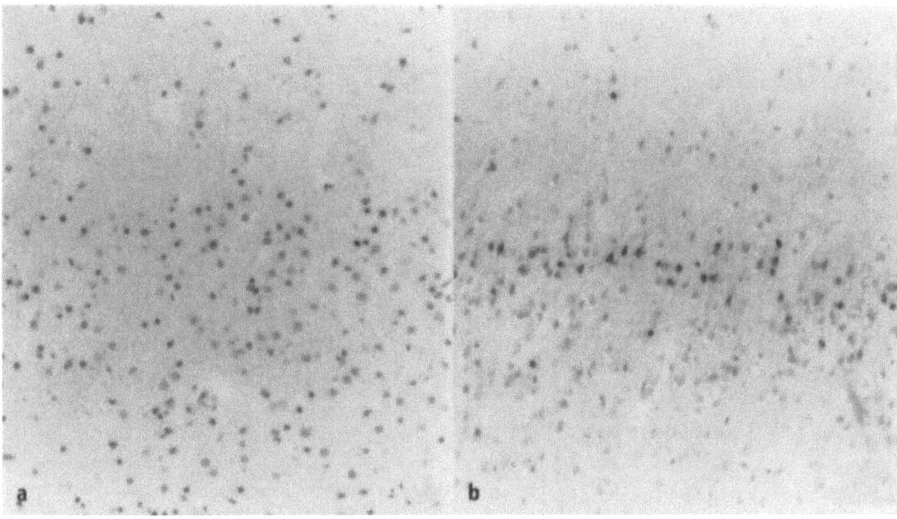

Fig. 3. Reduced immunoreactivity to Bcl-2 in the external granular layer of the temporal neocortex in DS subject (a), in comparison with normal control (b). 3-month old infants. x 200

consequence of enhanced cell death caused by insufficient protection of local neurons by Bcl-2. These findings indicate that a altered pattern of PCD might be, at least in part, responsible for abnormal cortical organization and subsequent aberrant brain functioning in DS subjects. However, the limited amount of DS brain tissues studied does not allow us to generalize this finding. In addition to reduced Bcl-2 immunoreactivity in the granular layer, in some DS brain tissues the pattern of Bcl-2 immunoreactivity was even at earlier stages of development more widespread, less organized and less consistent in the neocortex than that in normal human brain (Fig. 4). Moreover, although the hippocampal formation of the normal human brain showed only marginal, if any, immunoreactivity to Bcl-2 in the postnatal period, numerous Bcl-2-immunopositive neurons in the pyramidal layer of the hippocampus were discovered in one DS newborn. The pattern of Bax immunoreactivity in DS brain was rather similar to that in normal human brain. However, in some DS subjects, immunoreactivity to Bax was more widely distributed during both the gestational and postnatal periods of brain development. All these data suggest that altered PCD and, particularly, insufficient protection from apoptosis may contribute to brain pathology in DS subjects.

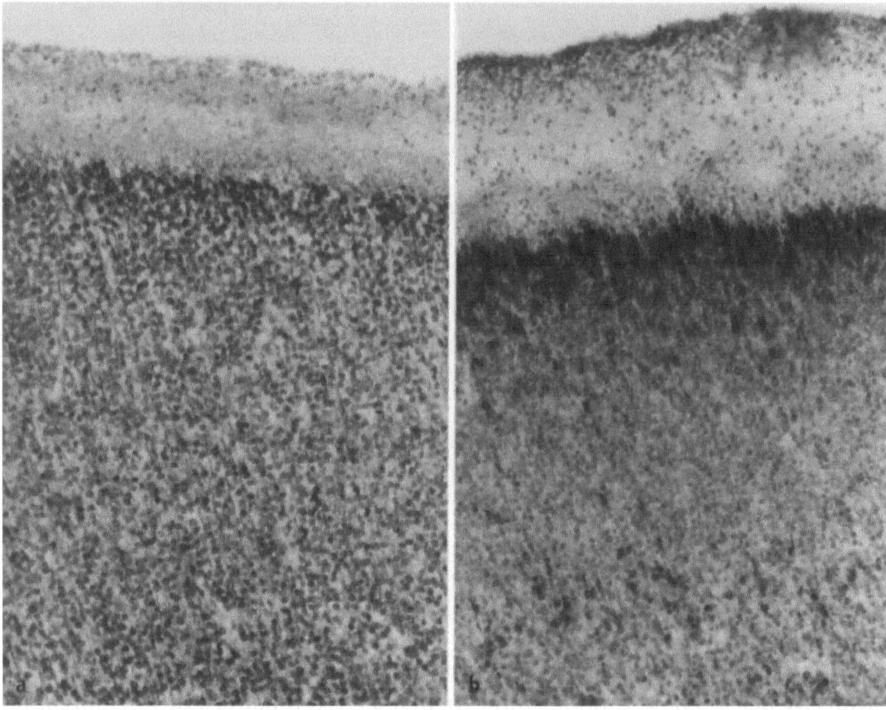

Fig. 4. More widespread and less organized pattern of Bcl-2 immunostaining of the temporal neocortex of DS (a) in comparison with normal fetal brain (b). Both at 20 wg. x 200

c-Jun in the Frontal and Temporal Cortex of the Normal and Down Syndrome Fetus

Recent data showed that some of the protooncogenes such as c-fos and c-Jun, which code for transcriptional factors, are implicated in PCD induced by growth factor deprivation in IL-2- and IL-6-dependent cell lines (Colotta et al. 1992). These genes belong to "immediate early genes" whose induction occurs within hour(s) after the initial stimulus. c-Jun induction in sympathetic neurons following nerve growth factor withdrawal is necessary for PCD to occur (Estus et al. 1994). c-Jun was found to be induced also in CNS neurons undergoing delayed cell death after ischemia (Dragunow et al. 1994). Increased immunoreactivity to c-Jun was recently described in cells with signs of PCD in neurodegenerative conditions such as AD (Anderson et al. 1996). However, according to our date, induction of c-Jun was not detectable by using immunocytochemical methods in developing human brain, even at the periods of intensive PCD. Only weak immunoreactivity to c-Jun, confined to a small number of cells, could be disclosed in both normal and DS brain. This observation suggests either that c-Jun induction during PCD lasted for too short a time to be detected or that natural PCD differs in this respect from experimentally induced PCD, which requires further study.

Concluding remarks

Our data support previous suggestions that the consequences of Bcl-2 family members interactions may depend upon the context in which they interact and the stage of development. The temporal and spatial patterns of Bcl-2 and Bax induction in the developing normal human brain indicate that both of these proteins, and especially Bax, may be involved not only in PCD but also in terminal differentiation of neurons in the human brain. Dissimilar topography and temporal distribution of these proteins in the normal control and DS brain indicate that PCD may be involved in the pathogenesis of DS brain abnormalities.

Acknowledgments

The authors wish to express their appreciation to Ms. Maureen Stoddard Marlow for editorial assistance, Mr. Lawrence Black for bibliographical assistance, and Ms. Jo-Ann Woodbridge Buttafuoco for secretarial assistance. This work was supported in part by funds from the New York State Office of Mental Retardation and Developmental Disabilities and a grant from the National Institute on Aging, Grant 5PO 1 AG11531-10.

References

Akao Y, Otsuki Y, Kataoka S, Ito Y, Tsujimoto Y (1994) Multiple subcellular localization of Bcl-2: detection in nuclear outer membrane, endoplasmic reticulum membranes and mitochondrial membranes. Cancer Res 54:2468-2471

Allsopp TE, Wyatt S, Paterson HF, Davies AM (1993) The proto-oncogene Bcl-2 can selectively rescue neurotrophic factor-dependent neurons from apoptosis. Cell 73:295-307

Anderson AJ, Su JH, Cotman CW (1996) DNA damage and apoptosis in Alzheimer's disease: colocalization with c-Jun immunoreactivity, relationship to brain area, and effect of postmortem delay. J Neurosci 16:1710-1719

Apert E (1914) Mongolism. Le Monde Medicale 24:201

Arends MJ, Wyllie AH (1991) Apoptosis: mechanisms and roles in pathology. Int Rev Exp Pathol 32:223-254

Baffy G, Miyashita T, Williamson JR, Reed JC (1993) Apoptosis induced by withdrawal of interleukin-3 (IL-3) from an IL-3-dependent hematopoietic cell line is associated with repartitioning of intracellular calcium and is blocked by enforced Bcl-2 oncoprotein production. J Biol Chem 268:6511-6519

Bakhshi A, Jensen JP, Goldman P, Wright JJ, McBride OW, Epstein AL, Korsmeyer SJ (1985) Cloning the chromosomal breakpoint of t(14; 18) human lymphomas: clustering around J_H on chromosome 14 and near a transcriptional unit on 18. Cell 41:899-906

Becker LE, Armstrong DL, Chan F (1986) Dendritic atrophy in children with Down's syndrome. Ann Neurol 20:520-526

Becker L, Mito T, Takashima S, Onodera K (1991) Growth and development of the brain in Down syndrome. Progr Clin Biol Res 373:133-152

Blaschke AJ, Staley K, Chun J (1996) Widespread programmed cell death in proliferative and postmitotic regions of the fetal cerebral cortex. Development 122:1165-1174

Brooksbank BWL, Balazs R (1984) Superoxide dismutase, glutathione peroxidase and lipoperoxidationin Down's syndrome fetal brain. Dev Brain Res 16:37-44

Brooksbank BWL, Martinze M, Balazs R (1985) Altered composition of polyunsaturated fatty acyl groups in phosphoglycerides of Down's syndrome fetal brain. J Neurochem 44:869–874

Brooksbank BWL, Walker D, Balazs R, Jørgensen OS (1989) Neuronal maturation in the fetal brain in Down's syndrome. Early Hum Dev 18:237–246

Busciglio J, Yankner BA (1995) Apoptosis and increased generation of reactive oxygen species in Down's syndrome neurons *in vitro*. Nature 378:776–779

Chittenden T, Harrington EA, O'Connor R, Flemington S, Lutz RJ, Evan GI, Guild BC (1995) Induction of apoptosis by the Bcl-2 homolog Bak. Nature 374:733–736

Claery ML, Sklar J (1985) Nucleotide sequence of a t(14; 18) chromosomal breakpoint in follicular lymphoma and demonstration of a breakpoint-cluster region near a transcriptionally active locus on chromosome 18. Proc Natl Acad Sci USA 82:7439–7443

Colon EJ (1972) The structure of the cerebral cortex in Down's syndrome: a quantitative analysis. Neuropädiatrie 3:362–376

Colotta F, Polentarutti N, Sironi M, Mantovani A (1992) Expression and involvement of c-fos and c-Jun protooncogenes in programmed cell death induced by growth factor deprivation in lymphoid cell lines. J Biol Chem 267:18278–18283

Cowan WM, Fawcett JW, O'Leary DD, Stanfield BB (1984) Regressive events in neurogenesis. Science 225:1258–1265

Crome L, Cowie V, Slater E (1996) A statistical note on cerebellar and brain stem weight in mongolism. J Ment Defic Res 10:69–72

Cronk CE (1978) Growth of children with Down's syndrome: birth to age 3 years. Pediatrics 61:564–568

Cronk CE, Crocker AC, Pueschel SM, Shea AM, Zackai E, Pickens G, Reed RB (1988) Growth charts for children with Down syndrome: 1 month to 18 years of age. Pediatrics 18:102–110

Davidoff LM (1928) The brain in mongolian idiocy. A report of ten cases. Arch Neurol Psych 20:1229–1257

Deckwerth TL, Elliott JL, Knudson CM, Johnson EM, Jr., Snider WD, Korsmeyer SJ (1996) BAX is required for neuronal death after trophic factor deprivation and during development. Neuron 17:401–411

Dragunow M, Beilharz E, Sirimanne E, Lawlor P, Williams C, Bravo R, Gluckman P (1994) Immediate-early gene protein expression in neurons undergoing delayed death, but not necrosis, following hypoxic-ischaemic injury to the young rat brain. Brain Res Mol Brain Res 25:19–33

Dubois-Dauphin M, Frankowski H, Tsujimoto Y, Huarte J, Martinou J-C (1994) Neonatal motorneurons overexpressing the Bcl-2 protooncogene in transgenic mice are protected from axotomy-induced cell death. Proc Natl Acad Sci USA 19:3309–3313

Ellis HM, Horvitz HR (1986) Genetic control of programmed cell death in the nematode. Cell 44:817–829

Ellis HM, Yuan JY, Horvitz HR (1991) Mechanisms and functions of cell death. Annu Rev Cell Biol 7:663–698

Estus S, Zaks WJ, Freeman RS, Gruda M, Bravo R, Johnson EM, Jr. (1994) Altered gene expression in neurons during programmed cell death: identification of c-Jun as necessary for neuronal apoptosis. J Cell Biol 127:1717–1727

Fernandes-Alnemri T, Litwack G, Alnemri ES (1994) CPP32, a novel human apoptotic protein with homology to *Caenorhabditis elegans* death protein Ced-3 and mammalian interleukin-1b-converting enzyme. J Biol Chem 269:30761–30764

Fitzsimmons J, Droste S, Shepard TH, Pascoe-Mason J, Fantel A (1990) Growth failure in second-trimester fetuses with trisomy 21. Teratology 42:337–345

Gadisseux J-F, Goffinet AM, Lyon G, Evrard P (1992) The human transient subpial granular layer: an optical, immunocytochemical, and ultrastructural analysis. J Comp Neurol 324:94–114

Galaburda AM, Kemper TL (1979) Down syndrome: is there a missing population of neurons? Neurology 29:569–570

Garcia I, Martinou I, Tsujimoto Y, Martinou J-C (1992) Prevention of programmed cell death of sympathetic neurons by the Bcl-2 proto-oncogene. Science 258:302–304

Golden JA, Hyman BT (1994) Development of the superior temporal neocortex is anomalous in trisomy 21. J Neuropathol Exp Neurol 52:513–520
Gonzáles-García M, Perez-Ballestero R, Ding L, Duan L, Boise LH, Thompson CB, Núñez G (1994) Bcl-x_L is the major Bcl-x mRNA form expressed during murine development and its product localizes to mitochondria. Development 120:3033–3042
Hara A, Iwai T, Niwa M, Uematsu T, Yoshimi N, Tanaka T, Mori H (1996) Immunohistochemical detection of Bax and Bcl-2 proteins in gerbil hippocampus following transient forebrain ischemia. Brain Res 711:249–253
Hengartner MO, Horvitz HR (1994) *C. elegans* cell survival gene ced-9 encodes a functional homolog of the mammalian proto-oncogene Bcl-2. Cell 76:665–676
Hockenbery DM, Zutter M, Hickey W, Nahm M, Korsmeyer SJ (1991) Bcl-2 protein is too topographically restricted in tissues characterized by apoptotic cell death. Proc Natl Acad Sci USA 88:6961–6965
Hockenbery DM, Oltvai ZN, Yin XM, Milliman CL, Korsmeyer SJ (1993) Bcl-2 functions in an antioxidant pathway to prevent apoptosis. Cell 75:241–251
Kemper TL (1988) Neuropathology of Down syndrome. In: Nadel L (ed) The psychobiology of Down syndrome. The MIT Press, Cambridge, pp 269–289
Kerr JFR, Wyllie AH, Currie AR (1972) Apoptosis: a basic biological phenomenon with wide-ranging implications in tissue kinetics. Brit J Cancer 26:239–257
Korsmeyer SJ (1995) Regulators of cel death. Trends Genet 11:101–105
Kostović I, Lukinowić N, Judaš M, Bogdanović N, Mrzljak L, Zečević N, Kubat M (1989) Structural basis of the development plasticity in the human cortex: the role of the transient subplate zone. Metab Brain Dis 4:17–23
Krajewski S, Krajewska M, Shabaik A, Miyashita T, Wang HG, Reed JC (1994) Immunohistochemical determination of in vivo distribution of Bax, a dominant inhibitor of Bcl-2. Am J Pathol 145:1323–1336
Lam M, Dubyak G, Chen L, Nuñez G, Miesfeld RL, Distelhorst CW (1994) Evidence that Bcl-2 represses apoptosis by regulating endoplasmic reticulum-associated Ca^{2+} fluxes. Proc Natl Acad Sci USA 91:6569–6573
Lee LC, Hunter JJ, Mujeeb A, Turck C, Parslow TG (1996) Evidence for α-helical conformation of an essential N-terminal region in the human Bcl-2 protein. J Biol Chem 271:23284–23288
Luskin MB, Shatz CJ (1985) Studies of the earliest generated cells of the cat's visual cortex: cogeneration of subplate and marginal zones. J Neurosci 5:1062–1075
Marin-Padilla M (1976) Pyramidal cell abnormalities in the motor cortex of a child with Down's syndrome. A Golgi study. J Comp Neurol 167:63–82
Marin-Padilla M (1984) Neurons of layer I: a developmental analysis. In: Peters A, Jones EG (eds) Cerebral cortex. Vol. I. Plenum Press, New York, pp 447–478
Martinou J-C, Dubois-Dauphin M, Staple JK, Frankowski H, Missoten M, Albertini P, Talabot D, Catsicas S, Pietra C, Huarte J (1994) Overexpression of Bcl-2 in transgenic mice protects neurons from naturally occurring cell death and experimental ischemia. Neuron 13:1017–1030
McConnell SK, Ghosh A, Shatz CJ (1989) Subplate neurons pioneer the first axon pathway from the cerebral cortex. Science 245:978–982
Merry DE, Veis DJ, Hickey WF, Korsmeyer SJ (1994) Bcl-2 protein expression is widespread in the developing nervous system and retained in the adult PNS. Development 120:301–311
Michaelidis TM, Sendtner M, Cooper JD, Airaksinen MS, Holtmann B, Meyer M, Toenen H (1996) Inactivation of Bcl-2 results in progressive degeneration of motoneurons, sympathetic and sensory neurons during early postnatal development. Neuron 17:75–89
Middleton G, Núñez G, Davies AM (1996) Bax promotes neuronal survival and antagonises the survival effects of neurotrophic factors. Development 122:695–701
Minn AJ, Boise LH, Thompson CB (1996) Bcl-x_S antagonizes the protective effects of Bcl-x_L. J Biol Chem 271:6306–6312
Monaghan P, Robertson D, Amos TA, Dyer MJS, Mason DY, Greaves MF (1992) Ultrastructural localization of Bcl-2 protein. J Histochem Cytochem 40:1819–1825

Mu X, He J, Anderson DW, Trojanowski JQ, Springer JE (1996) Altered expression of Bcl-2 and Bax mRNA in amyotrophic lateral sclerosis spinal cord motor neurons. Ann Neurol 40:379–386
Núñez G, London L, Hockenbery D, Alexander M, McKearn JP, Korsmeyer SJ (1990) Deregulated Bcl-2 gene expression selectively prolongs survival of growth factor-deprived hematopoietic cell lines. J Immunol 144:3602–3610
Oltvai ZN, Korsmeyer SJ (1994) Checkpoints of dueling dimers foil wealth wishes. Cell 79:189–192
Oltvai ZN, Milliman CL, Korsmeyer SJ (1993) Bcl-2 heterodimerizes in vitro with a conserved homolog, Bax, that accelerates programmed cell death. Cell 74:609–619
Petit TL, LeBoutillier JC, Alfano DP, Becker LE (1984) Synaptic development in the human fetus: a morphometric analysis of normal and Down's syndrome neocortex. Expl Neurol 83:13–23
Rabinowicz T, de Courten-Myers GM, Mc-Donald-Comber Petetot J, Xi G, De Los Reyes E (1996) Human cortex development: estimates of neuronal numbers indicate loss late during gestation. J Neuropathol Exp Neurol 55:320–328
Ross MH, Galaburda AM, Kemper TL (1984) Down's syndrome: is there a decreased population of neurons? Neurology 34:909–916
Sara VR (1986) Brain development. In: Wortis J (ed) Mental retardation and developmental disabilities. Vol. 14. Elsevier, New York, pp 163–181
Sara VR, Gustavson KH, Anneren G, Hall K, Wetterberg L (1983) Somatomedins in Down's syndrome. Biol Psych 18:803–811
Satou T, Cummings BJ, Cotman CW (1995) Immunoreactivity for Bcl-2 protein within neurons in the Alzheimer's disease brain increases with disease severity. Brain Res 697:35–43
Schmidt-Sidor B, Wisniewski KE, Shepard TH, Sersen EA (1990) Brain growth in Down syndrome subjects 15 to 22 weeks of gestational age and birth to 60 months. Clin Neuropathol 9:181–190
Schwartz LM, Smith SW, Jones MEE, Osborne BA (1993) Do all programmed cell deaths occur via apoptosis? Proc Natl Acad Sci USA 90:980–984
Scott BS, Petit TL, Becker LE, Edwards BAV (1982) Abnormal electric membrane properties of Down's syndrome DRG neurons in cell culture. Dev Brain Res 2:257–270
Sentman CL, Shutter JR, Hockenbery D, Kanagawa O, Korsmeyer SJ (1991) Bcl-2 inhibits multiple forms of apoptosis but not negative selection in thymocytes. Cell 67:879–888
Steinman HM (1995) The Bcl-2 oncoprotein functions as a pro-oxidant. J Biol Chem 270:3487–3490
Suetsugu M, Mahraein P (1980) Spine distribution along the apical dendrites of the pyramidal neuron in Down's syndrome. A quantitative Golgi study. Acta Neurpathol (Berl) 50:207–210
Takashima S, Becker LE, Armstrong DL, Chan F (1981) Abnormal neuronal development in the visual cortex of the human fetus and infant with Down's syndrome. A quantitative and qualitative Golgi study. Brain Res 225:1–21
Tewari M, Dixit VM (1995) Fas- and tumor necrosis factor-induced apoptosis is inhibited by the poxvirus crmA gene product. J Biol Chem 270:3255–3260
Thornberry NA, Bull HG, Calaycay JR, Chapman KT, Howard AD, Kostura MJ, Miller DK, Molineaux SO, Weidner JR, Aunins J, Elliston KO, Ayala JM, Casano FJ, Chin J, Ding GJ-F, Egger LA, Gaffney EP, Limjuco G, Palyha OC, Raju SM, Rolando AM, Salley JP, Yamin T-T, Lee TD, Shively JE, MacCross M, Mumford RA, Schmidt JA, Tocci MJ (1992) A novel heterodimeric cysteine protease is required for interleukin-1b processing in monocytes. Nature 356:768–774
Tiso N, Pallavicini A, Muraro T, Zimbello R, Apolloni E, Valle G, Lanfranchi G, Danieli GA (1996) Chromosomal localization of the human genes, CPP32, Mch2, Mch3, and Ich-1 involved in cellular apoptosis. Biochem Biophys Res Commun 225:983–989
Tsujimoto Y, Gorham J, Cossman J, Jaffe E, Croce CM (1985) The t(14;18) chromosome translocations involved in B-cell neoplasms result from mistakes in VDJ joining. Science 229:1390–1393
Veiš DJ, Sorenson CM, Shutter JR, Korsmeyer SJ (1993) Bcl-2-deficient mice demonstrate fulminant lymphoid apoptosis, polycystic kidneys and hypopigmented hair. Cell 75:229–240
Veis Novack D, Korsmeyer SJ (1994) Bcl-2 protein expression during murine development. Am J Pathol 145:61–73
Vito P, Lacaná E, D'Adamio L (1996) Interfering with apoptosis: Ca^{2+}-binding protein ALG-2 and Alzheimer's disease gene ALG-3. Science 271:521–525
Volpe JJ (1977) Normal and abnormal human brain development. Clin Perinatol 4:3–30

Wang L, Miura M, Bergeron L, Zhu H, Yuan J (1994) Ich-1, an ICE/ced-3-related gene, encodes both positive and negative regulators of programmed cell death. Cell 78:739–750

Wisniewski KE (1990) Down syndrome children often have brain with maturation delay, retardation of growth, and cortical dysgenesis. Am J Med Genet Suppl 7:274–281

Wisniewski KE, Kida E (1994) Abnormal neurogenesis and synaptogenesis in Down syndrome brain. Dev Brain Dysfunct 7:289–301

Wisniewski KE, Laure-Kamionowska M, Wisniewski HM (1984) Evidence of arrest of neurogenesis and synaptogenesis in brains of patients with Down's syndrome (letter). N Engl J Med 311:1187–1188

Wyllie AH, Kerr JFR, Curri AR (1980) Cell death: the significance of apoptosis. Int Rev Cytol 68:251–306

Yaginuma H, Tomita M, Takashita M, McKay SE, Cardwell C, Yin Q-W, Oppenheim RW (1996) A novel type of programmed cell death in the cervical spinal cord of the chick embryo. J Neurosci 16:3685–3703

Yang E, Zha J, Jockel J, Boise LH, Thompson CB, Korsmeyer SJ (1995) Bad, a heterodimeric partner for Bcl-x_L and Bcl-2, displaces Bax and promotes cell death. Cell 80:285–291

Yin X-M, Oltvai ZN, Korsmeyer SJ (1994) BH1 and BH2 domains of Bcl-2 are required for inhibition of apoptosis and heterodimerization with Bax. Nature 369:321–323

Yuan J, Shaham S, Ledoux S, Ellis HM, Horvitz HR (1993) The C. elegans cell death gene ced-3 encodes a protein similar to mammalian interleukin-1-b-converting enzyme. Cell 75:641–652

Zhong L-T, Sarafian T, Kane DJ, Charles AC, Mah SP, Edwards RH, Bredeen DE (1993) Bcl-2 inhibits death of central neural cells induced by multiple agents. Proc Natl Acad Sci USA 90:4533–4537

Zhu W, Cowie A, Wasfy GW, Penn LZ, Leber B, Andrews DW (1996) Bcl-2 mutants with restricted subcellular location reveal spatially distinct pathways for apoptosis in different cell types. EMBO J 15:4130–4141

Subject Index

Springer and the environment

At Springer we firmly believe that an international science publisher has a special obligation to the environment, and our corporate policies consistently reflect this conviction.
We also expect our business partners – paper mills, printers, packaging manufacturers, etc. – to commit themselves to using materials and production processes that do not harm the environment. The paper in this book is made from low- or no-chlorine pulp and is acid free, in conformance with international standards for paper permanency.

Springer

MIX
Papier aus verantwortungsvollen Quellen
Paper from responsible sources
FSC® C105338

If you have any concerns about our products,
you can contact us on
ProductSafety@springernature.com

In case Publisher is established outside the EU,
the EU authorized representative is:
Springer Nature Customer Service Center GmbH
Europaplatz 3, 69115 Heidelberg, Germany

Printed by Libri Plureos GmbH
in Hamburg, Germany